The Environment

Our Common Heritage

Edited by
Richard W. Bell
Jimmy Seow

AF677456

ASSOCIATION FOR BAHÁ'Í STUDIES AUSTRALIA
IN ASSOCIATION WITH
BAHÁ'Í PUBLICATIONS AUSTRALIA

© Association for Bahá'í Studies Australia
P.O. Box 285
Mona Vale NSW 2103
Australia
ISBN 0 909991 70 7

First Edition 1994

This book is based on Papers presented at the 9th Australian Baha'i Studies Conference, held at Murdoch University on Saturday, 14 April, 1990.
The views expressed in these papers are those of the contributors and are not necessarily those of the Association for Baha'i Studies or the National Spiritual Assembly of the Baha'is of Australia, Inc.

Monograph No. 5

Printed in Australia

Contents

“We cannot segregate the human heart from the environment outside us and say that once one of these is reformed everything will be improved. Man is organic with the world. His inner life moulds the environment and is itself also deeply affected by it. The one acts upon the other and every abiding change in the life of man is the result of these mutual reactions."

Shoghi Effendi - Bahá'í Writings

OPENING ADDRESS

Professor Peter Boyce [1]

Mr Chairman, ladies and gentlemen, I extend a very warm welcome to participants to this most important conference. I hope you will agree that Murdoch is an appropriate venue for such a conference, not just because we host a strong environmental science programme but also for the possibly less immediately obvious reason that this University has been one of the first in Australia to encourage holistic approaches to learning in all schools, which I understand to be a very central tenet of Bahá'í teachings about the environment and humankind's relationship to it. As a Christian, I also feel that the choice of Easter Saturday for a conference on this theme was not at all inappropriate, though I confess to having harboured initial doubts about it.

Environmentalists as an organized political force have never wielded so much clout in Western Industrialized societies as today. But analyses by political journalists and by number- crunching consultants to the major political parties tell only half the story, possibly a misleading one at that. For it is surely not in the long term interest of the environment that caring citizens be identified solely as a minority interest group or political pressure group. Subtle and gradual erosion of old values are taking place across the political spectrum, and that surely is the only kind of erosion environmentalists are likely to applaud.

Australia has more reason than most to work for rapid change in both public policy and private attitudes. A country which carries the inherent disadvantage of being the world's driest continent and which nevertheless managed to destroy half of its forests and one third of its

---

1 Vice-Chanellor, Murdoch University

woodland and scrub within the first one hundred and fifty years of European settlement has quite literally a lot of ground to make up.

Nevertheless, in seeking to change public policy and private attitudes there needs to be a convergence of arguments for the environment with arguments for economic growth. And I want to reflect for a couple of minutes on an impressive paper presented by a Federal Cabinet Minister in the lead-up to the last election. I hold no brief for either the party or the Minister, but I do feel that John Kerin, as Minister for Primary Industry and Energy at that time, was making a very serious attempt to link two basically inextricable issues in Australian public debate. Given that it was an address to the Committee for Economic Development, the sympathetic approach to environmental problems which he revealed is all the more creditable.

John Kerin said that there were five basic propositions which everyone had to address or consider before there could be any meaningful follow-up discussion or action concerning the environment, particularly Australian environmental problems. First of all, he said, the use of a non-renewable resource implies the eventual exhaustion of that resource. Secondly, the capacity of this country to meet the increasing demand for the maintenance of our environment depends on continued increases in national income. Now we may wish to challenge that. And I will come back to that in a moment. The third one: the community is entitled to an appropriate social return from exploitation of the resources vested with the community. Fourthly, development of a consistent national resource policy requires a common accord between the States and the Commonwealth. Finally, there exists a huge chasm between attitudinal change and behavioural change.

And these are just propositions about the Australian scene. Of course, additional propositions could be made about the global picture. For example, the durability of the nation state in international relations and the difficulties of finding, and then imposing, suitable sanctions on delinquent states must never be lost sight of by domestic lobby groups concerned with environmental policies. But if I return to Kerin's five propositions, the first one, about the exhaustion of non-renewable resources, is probably incontestable. The one on economic growth is more difficult. What Kerin was saying is that negative economic growth does not provide an answer to legitimate environmental concerns. A one-off decline in gross national product, say five percent over two years, Kerin argues, would lead to the following:- A decline in real household consumption by 0.8 percent, a decline of national savings by 22.6 percent, a decline in aggregate investment by 18.6 percent and a decline in employment by 1.5 percent. I have no way of checking on the accuracy of these figures but I think he's making a very persuasive proposition. If, however, the community at large had agreed on a collective self-denial ordinance concerning their standard of living, Mr. Kerin would probably not need to be so concerned about the primacy of economic growth. But that is not likely to happen, at least in the short term.

Kerin's third proposition, that there needs to be an appropriate social return, is really taking up the question of appropriate taxation levels on resource industries, and he says that the taxation must be efficient, consistent and equitable. Rents belong to the community. He is also making the important point that external costs are not accounted for on a company's balance sheet. And this, I suppose, is the most contentious problem in resource management. The monetary value of clean air or pure water in native forests is not easily calculated. It certainly does not appear on a company's balance sheet.

The fourth proposition that the Minister offered, about the need for an accord between the Commonwealth and the States, has been pursued by political scientists and constitutional lawyers for some time but not necessarily around the theme of environmental protection. Kerin quite rightly says as a centralist, the states are skittish by nature. Mr Whitlam as a Labor Prime Minister some ten years before said the States were "an abomination". Well, one cannot expect "State's righters" or State political parties to agree with that observation but certainly in the world of environmental problems a Minister at the central level of government is entitled to feel frustrated. Kerin concedes that the States must be gently wooed and then bribed. Of course, he would like to see resource powers transferred to the Commonwealth. My own prediction, and I think he would agree, is that if a referendum were held tomorrow on this question, the Commonwealth would lose. Nevertheless, many of us would probably see this as a desirable long-term objective.

When we come to the question of attitudinal change, we are taking up the theme of education for the environment. Too many of our brightest brains in this country are being encouraged to pursue profit-making activity, such as the money market, rather than employment which focuses on urgent social and environmental issues. Murdoch's School of Biological and Environmental Science has trained many graduates over the last fifteen years for useful careers that will not necessarily earn huge incomes. And I am pleased to report a dramatic increase in enrolments in Environmental Science in recent years.

We all know that the basic problem is human greed and conspicuous consumption. We should also recognize that there are limitations inherent in the liberal democratic system of government for those wishing to pursue environmental reform. Governments do not like taking long term precautions with long term planning. They not only cost money; they do not win votes. Governments are by definition very much preoccupied with short term interests and limited planning. Nor, I would have to say, can there be any early realization of an international parliament, though I acknowledge that in Bahá'í teaching a very high priority is given to the ideal of an international government. Nevertheless, international treaties, both bilateral and multilateral, covering protection of the environment are increasing in nature and scope, while the education of international opinion (and there is now a phenomenon called international opinion) continues apace. These areas of activity are yielding results: so I hope that Bahá'í colleagues here will not feel that the absence of an international parliament will necessarily prevent there being considerable progress on the international front towards protection of the environment.

These are attitudinal problems and educational problems, and I can assure you that at Murdoch University we will be attaching very considerable importance to the educational dimension. This all points, it seem to me, to the need for more not less government regulation. And that of course sets us apart from those arguing for a deregulation on many fronts. The environmental catastrophies of oil spills, acid rain, atmospheric pollution, deforestation and noxious wastes provide a very large urgent agenda for strong public measures to be taken. That does not necessarily counter the argument for small government. But it does counter the argument for deregulation.

A recent paper by Peter Self from London University and the Australian National University on public choice reminds us that more than half of the questions raised in the European parliament concern deficiencies of public regulation most of them concerning environmental protection.

And so we get back to this need for holistic approaches to life and to protection of the environment. I believe that the mainstream Christian churches have let us down badly here, notwithstanding the invitation to an holistic approach which is carried through the scriptures. In practice, in week by week preaching, in the daily message which issues from the mainstream Christian churches there is far too little evidence of a recognition of environmental concerns and too little attempt to adopt an holistic approach to life.

There are some encouraging news items. Within this country, the fact that the government has declared 1990 as the start of a Decade of Landcare, with several hundred million dollars committed to soil conservation and the planting of trees, is encouraging, although the national facilitator on land care, Andrew Campbell, has written that the money, $47m, allocated toward the planting of a billion trees is only about 10 percent of what is needed.

There is too a growing unity of purpose between farmers and graziers on the one hand, and environmentalists on the other. We find also more extensive media coverage of environmental problems. And there is a higher salience of environmental issues in political party platforms, while in the State of Tasmania until recently, the environmentalists actually held the balance of power.

Internationally, there are also some signs of progress. I read recently that the Netherlands has adopted a 20 year environment plan. And there may be other Western European countries about to do the same. Action is now possible in tackling the environment in Eastern Europe following the political liberation of so many countries there. There are more and more multilateral agreements covering environmental problems, ranging from Antarctica to the dumping of nuclear wastes, to the Montreal Protocol setting out a phase out schedule for the chlorofluorocarbons (and I note that the Australian Conservation Foundation says there has been a significant improvement in the understanding of the problems of global atmospheric ozone depletion since the adoption of that Montreal Protocol in 1987).

I have said enough, ladies and gentlemen, by way of a personal introduction to this vital theme, possibly too much, considering that I am a layman. May you emerge from these proceedings late this afternoon both inspired and instructed. I have very great pleasure in formally declaring this conference open.

PREFACE

Jimmy Seow[1]

As we enter the closing decade of this century, humanity is confronted with many vital issues it must address and resolve. Issues which are both challenging and perplexing not only to the leaders of the world but also to the common person in the street. Some of these issues are political in nature concerning the attainment of peace and establishment of a more stabilised world. Other issues are social and moral as we are forced to deal with the current flux of mass migration and settlement of refugees. The striving for a more equitable and just society through the eradication of abject poverty and the narrowing of the widening gap between the haves and have-nots is another important issue. There are many other important issues which time does not permit me to list but one issue which is rapidly assuming world-wide importance in both the Developed and Third World countries is the question of the quality of our environment - humanity's common heritage. Questions have been asked and are still being asked regarding what kind of world we want to live in. Mother Earth, our common habitat can only take so much abuse and pollution. One is fully aware of the present environmental abuses taking place - the indiscriminate clearing of the remaining forests; the unabating chemical pollution of our biosphere and atmosphere; the alteration of the earth's climatic and weather patterns; and the extinction of species of both fauna and flora due to man's incessant appetite for material consumption. These are some of the environmental issues confronting us. We are the caretakers for future generations and we must act responsibly. In the words of Bahá'u'lláh, the Founder of the Bahá'í Faith, "All men have been created

---

1 Senior Environment Officer, Petroleum Division, Mines Department, W.A.

to carry forward an ever-advancing civilization."[2] Norman Myers points out the need for a new international perspective to address these environmental problems confronting humankind. He stated that "many emergent forms of environmental impoverishment and resource degradation constitute a distinctive category of international problems, unlike anything in the past. In this instance, economic well-being, political stability and overall security have become to be recognised as indivisible - as indivisible as the biosphere. To this extent, environmentalism helps to articulate the shared interest of nations: confrontation gives way to co-operation."[3]

Humankind is an integral part of the environment. Our harmonious relationship with other living organisms that share a common habitat is quintessential for our existence. This sentiment is also expressed in the Bahá'í Writings that "Man is organic with the world. His inner life moulds the environment and is itself also deeply affected by it. The one acts upon the other and every abiding change in the life of man is the result of these mutual reactions."[4] The onus is therefore placed upon us to ensure that there will be a future for generations to come. Hence we need to, using the phrase of the world famous scientist/environmentalist, Dr. David Suzuki, "invent the future".[5]

What then constitutes the environment? Why is the preservation and conservation of the environment so important and crucial? What is our role in its preservation? These are some of the questions for which we sought answers. To this end, the Association for Bahá'í Studies in Australia invited speakers from various professions, religions and persuasions of thought to present their thoughts, understanding and ideas on

---

2 Bahá'u'lláh, Gleanings from the Writings of Bahá'u'lláh, Wilmette: Bahá'í Publishing Trust, 1952, CIX, p.215.

3 Norman Myers, "Environment and Security", The Environmentalist, December 1986.

4 Shoghi Effendi, in: A Bahá'í Perspective on Nature and the Environment, Mona Vale: Bahá'í Publications Australia, 1986, pp.9-10.

5 David Suzuki, Inventing the Future, Sydney: Allen & Unwin, 1990.

this important issue - the environment, firstly at a Symposium held at Murdoch University on 14 April 1990 and then subsequently by writing up their papers for publication in the present monograph. I trust through their papers and the ensuing dialogue, horizons of understanding will be further broadened and deepened.

I would like to express my appreciation to the services rendered by Mick Cater for his assistance in recording the proceedings of the symposium, and Mini Tan for typing and transcribing several of the papers. Their services greatly contributed to the completion of this publication.

OUR ENVIRONMENT - PAST, PRESENT AND FUTURE

Peter Newman[1]

Introduction

In 1975 I started the first course in Environmental Science at Murdoch University. It was one of the first Universities to make a commitment to this all-embracing discipline. For most academics and community leaders at the time it was considered to be a trendy fashion that would soon go away; it was certainly not something any sensible undergraduate would want to take as they would never get a job in the real world of economists and engineers.

Today our graduates are eagerly sought, the economists and engineers are looking for advice, decision-makers are worried whether they are green enough and the media phenomenon goes on getting bigger. It has been a big change.

Global awareness about environmental issues today is huge. For many people awareness was considered to be the only important factor. Once there was enough awareness "they" would act to stop all this pollution etc. "They" was always a bit undefined but in general it was a mixture of government and industry. But of course "they" also includes "us". And when we see the extent of our environmental problems and the long history that lies behind them, then we realise that it is not just a simple awareness issue like a Public-Relations campaign. What we are looking at is a fundamental issue concerning the whole nature of human society and of the human condition. At this time it is good to pause and think about these deeper aspects of the environmental problem.

---

1 Associate Professor in Environmental Science and Director, Institute for Science and Technology Policy, Murdoch University, Perth, W.A..

Global Environmental Issues

What is our present status on environmental matters, particularly in the global context because this is one issue on which we can not afford to be only local?

I have summarised it into 10 points:

1. **Rain Forest Deterioration**

We hear of destruction of the rain forests; a tragedy not only because of the loss of species but these forests are a vital part of the global climate. We are losing them mainly through burning but also from timber and woodchips for the western world. The extent of the destruction has been described by the International Tropical Timber Organisation.[2]

- Only 1 per cent of the world's 800 million hectares of productive tropical forests are being managed without irreversible damage.
- The world has already lost half of the 1.5 billion hectares of humid forest with which it began this century.
- Between 10 and 20 million hectares of rainforest are being cleared or burnt each year. One aerial photo showed 30,000 fires alight simultaneously in the Amazon rain forest.

2. **Global Population**

We hear about a global population that has now reached 5 billion and is moving quickly towards 6 billion - perhaps stabilising by 2050 at 10 billion.

According to Jim McNeil from the UN Bruntland Commission: "Every 10 seconds the world's population grows by 25 people".[3]

---

2 Anonymous, "Grim prospects for rainforest", *Our Planet* 1 (3) 1989, pp. 12-13.

3 World Commission on Environment and Development, Our Common

3. **Land Degradation**

We hear of land degradation, loss of soil, increases in salinity and desertification occurring to such an extent that 12 million acres of land globally each year is lost from production into worthless desert.[4] That represents the loss of 23 acres every minute. In Western Australia (WA) saline land continues to grow by doubling every 14 years.[5]

4. **Fish Loss, Pollution of Water**

We hear of fishery stocks depleting and species vanishing from the oceans, of water pollution in the ocean, streams and rivers and ground water, and the continuing loss of wetlands. WA's estuaries and wetlands are where we see our most important natural ecosystems and everywhere they are under threat.

5. **Chemical Industry Failure**

We hear of Bhopal where 2000 people were killed by a pesticide factory explosion and 200,000 were injured, and of Chernobyl where radiation spread across the earth showing the awesome potential of industry failure.

6. **City Growth and Urban Sprawl**

We hear about cities growing bigger and bigger, becoming more and more dominated by cars and freeways and sprawling out into farms and bush. And in Third World cities the shanty towns are growing faster than anywhere else. City sprawl is our biggest urban problem in Perth as the city spreads into the hills, wetlands, and bush at a rate apparently beyond control.

7. **Greenhouse and Ozone Depletion**

Future, Oxford University Press: Oxford, 1987.

4 World Commission on Environment and Development, ibid.

5 R.J. George, "1989 Saltland Census", Journal of Agriculture Western Australia, Series 4, 31, 1990.

We hear of the Greenhouse Effect where the burning of fossil fuels and forests threatens the climate we assume provides for all our life's certainties of food supply and water.

Australian cities are some of the highest gasoline users in the world and Perth leads the way in Australia.[6]

We hear of ozone depletion due to chlorofluorocarbons (CFC's) breaking a hole in the very sky above our heads and threatening a skin cancer epidemic.

8. Abject Poverty in the Third World

We hear of those parts of the world where there is abject poverty, where basic needs are not being provided, where there is suffering and little human dignity.

As John Langmore, the President of ALP Third World Forum says: "There are now nearly a **billion people** living in absolute poverty, a condition of life so degraded by disease, illiteracy, malnutrition and squalor as to deny its victims basic human necessities."[7] Such a domestic environment means water supply and waste disposal are a daily risk.

9. Environmental Refugees

We hear of the huge population of displaced people. In many parts of Africa more than two percent of the population have become "environmental refugees" - displaced by land degradation.

10. Global Arms Race and Threat of Nuclear War

Despite recent understandings between the superpowers we hear

---

6 P. W. G. Newman and J. R. Kenworthy, Cities and Automobile Dependence: An International Sourcebook, Aldershot: Gower, England, 1989.

7 J. Langmore and D. Peetz (Eds.), Wealth, poverty and survival: Australia in the world, Sydney: George Allen & Unwin, in association with the Australian Labour Party, 1990.

of the huge global arsenals with their ability to destroy us hundreds of times over.

It's a pretty depressing list - but its basic message of doom is not new. The list is best summarized by St Paul 2000 years ago who said:

"The whole world groans as though in childbirth" [8]

It is worth remembering that environmental problems did not start in the 1970's.

Environmental History

I have a book by George Marsh written in 1860's about the United States of America.[9] It reads like a 1980's text on the problems of cutting forests and overcultivation of land.

I have another book by Fairfield Osborn called "Our Plundered Planet" which systematically goes through the great civilisations and shows how they destroyed their environment and thence led to their own destruction.[10] For example . . .

Greece was once 60% covered by fine forests. The cutting for cities and overgrazing denuded their hills. Now there is less than 2% of Greece's original soil left.

Turkey was also once forested. Now the once great port city of Tarsus lies 10 miles inland as the river has carried out so much eroded soil.

The **Cedars of Lebanon** were once the greatest living natural resource of the Middle East - now there are 12 left on the slopes of Mt Lebanon.

---

8 New Testament, Romans Chapter 8 Verse 22.

9 G. P. Marsh, Man and Nature, 1864. Reprinted Oxford University Press: Oxford, 1965.

10 F. Osborn, Our Plundered Planet, Little, Brown and Company: Boston, 1948.

Dead cities litter the **North of Africa**, once the wheatbelt of the Roman Empire.

China and India have huge areas of denuded and eroded land which was once forested. The Yellow River now runs through embankments above the level of the valley as it is so full of silt.

Every place humans touch seems to have some negative impact on the environment.

We cannot even romanticise the Australian Aboriginal - their use of fire seems to have had a devastating effect on flora and fauna, clearing out much of the closed wet forest for dry open woodland. There is even some recent evidence that the coming of Aborigines to Australia and their use of fire induced a climate change.[11]

But is it all really that depressing? Can we do nothing good at all? How do we begin to understand our history and our role in managing the environment?

I have another book which helps on these matters - the Bible. As I am an amateur theologian I would like to give a little Judeo-Christian background, because I know something about that, though I am sure similar perspectives can be found in most of the scriptures of the world's religions.

Environmental Theology

Genesis (a book of stories that go deep into prehistory and oral traditions) starts with a creation that is declared by its Creator to be "good" - its diversity and beauty are fully recognised. Man is given a role, his first function actually, to be an ecologist and name everything, and then to manage creation. There is harmony in the Garden of Eden but a fatal choice is made and humans are ejected never to return. Whatever else it means we do seem to have a faint memory of a perfect

11 A.P. Kershaw, "Was there a Great Australian Arid Period?" Search, 20 (3), 89, 1989.

natural order and dream to be part of it - utopias are still being created, but they always fail. There is no noble savage, no romantic harmonious paradise untouched by the evils of human civilization. There is no way back to Eden. I read a story once about the collapse of an ecologically based hippy commune called "Return to Eden" which also came to this conclusion.

The history of the Old Testament prophets covers the various civilisations that rose and fell over several thousand years. Their approach was essentially to describe two cities - **Babylon and Zion** - two cities of the mind that represent the two sides of human potential. These images are the key that I believe helps us to understand the spiritual aspects of the environmental issues of our past, present and future. So let me tell you a little about these two city images - cities that exist on a spiritual plane but which give substance to the physical cities in which we live.

Babylon was the city that lived only for itself, the city that in its arrogance exploited people and nature without concern for the future. As Isaiah said:

> "You thought in your own mind I will scale the heavens
> and make myself like the Most High . . . Babylon - the
> man who shook the earth who turned the world into a
> desert, who never let his prisoners go free to their
> homes." [12]

Babylon of course was an actual city that collapsed, due in part to its arrogant abuse of the forests in the surrounding mountains which it stripped for its war machine, with the inevitable soil erosion leading to silting of the Euphrates and the collapse of the irrigation system on which Babylon depended. Isaiah hints at this when he saw a vision of how the oppressor would meet his end, and then:

> "The whole world has rest and is at peace it breaks into

12 Old Testament, Isaiah, Chapter 14, Verse 13.

cries of joy. The pines themselves and the Cedars of
Lebanon exult over you, Since you have been laid low
they say: No man comes up to fell us."[13]

This raises a very important point for our understanding of environmental matters. History was never inevitable, there was always alternative visions from the prophets of old. Isaiah had visions of doom but also visions of hope.

The image of Babylon (the City of Man) was always in the context of Zion (the City of God).

"What answer is there for the envoys of the nation?" asks Isaiah, "This, that the Lord has fixed Zion in her place, and the afflicted among his people shall take refuge there."[14]

This place was where "the lion will lie down with the lamb", where peace and justice existed, where there was "milk and honey" (i.e. both agriculture and wilderness were in harmony with humanity).

For every message of despair and doom from the prophets there was a message of hope. To leave out the hope is to miss the distinctive message of all faiths - then and now. I will therefore return to give some hope but first let me pursue these alternative images of the future.

The images of Babylon and Zion are continued throughout the Bible long after Babylon as a place had become a wasteland. They are important spiritual images.

Revelation, the last book in the Bible is very difficult to understand. It sets out a series of images which many great theologians have tried to unravel and to which I will not be adding any great insights. But it is easy to get lost in the details of 144,000 of this and 7 of that when the central images are of Babylon and Zion - competing futures.

BABYLON is presented as the great city which has no values other than its own glory. It represses and exploits and delights in its

13 Old Testament, Isaiah, Chapter 14, Verses 7-8.
14 Old Testament, Isaiah, Chapter 14, Verse 32.

wealth and finery and power. But it is under judgement and this comes in several images - seven angels are described which bring the judgements:

- The first depicts people with malignant sores.
- The second has the oceans polluted blood red and dying (red tides are an algae that occurs in polluted waters).
- The third has rivers and springs polluted and dying in the same way.
- The fourth has the sun burning people dreadfully (an ozone layer image maybe?)
- The fifth has darkness, like an energy crisis.
- The sixth has rivers drying up (an image of climate change?) followed by plagues.
- The seventh has an earthquake and fire image that destroys everything (not unlike what a nuclear war could do).

This is Babylon - this is all that is wrong, all that can be if we continue in one particular direction. It is an image we know so well as it has been graphically shown to us on the television programmes describing our environmental crises and how they could develop. Its essential characteristic is that it lives for itself - such values give it power but lead to its end - it sees no-one else, it sees no environmental constraint, or any other kind of limit on its operations.

ZION - the heavenly city, the city that is made by God. What are its characteristics? What is the alternative vision to that of Babylon?

For a start it is a city - it is not the Garden of Eden revived. It is not a romantic medieval rustic village or a hunter and gatherer primitive culture. It does not turn its back on human technology and ingenuity - it perfects it.

It is a city made of "priceless jewels" and with "streets of gold".

Jewels and gold are, of course, dug out of the ground and refined with human skills. All of the technical skills of the scientist, engineer and artist are combined when it comes to cutting diamonds and other jewels. So the city of Zion expresses all that is best in human endeavour.

It is also a city with a tree of life and a river of life, yielding productive things. There is life in the city in harmony with human endeavour. The human environment encompasses the natural, there is harmony between the people and there is life.

Such is the alternative vision - the vision of hope. So what do these alternative images mean for us in the 1990's?

Application of Alternative Visions

The questions we have to ask ourselves are:

- Is the vision of Babylon that we hear described so often really inevitable?
- Are we just bystanders in this world as we watch the cosmic drama unfold?
- Are we just victims of history, especially a history of ecological disaster?
- Is it just a question of awareness to solve these problems?

I would want to answer all these questions with a resounding **NO!**

- The visions of Babylon presented daily to us on the TV are not inevitable . . . The Bible and history is full of stories where judgement was diverted by human society changing its priorities and values.
- We are not just bystanders in a cosmic drama. We have choice, we have a purpose and role in life, we are not just puppets.
- Awareness is not enough though it is a precursor. Solutions involve choices which are essentially part of a spiritual

battle and that should never be underestimated. It will be a battle in our personal choices.

- But our work and our lives become meaningful in the light of the alternative future directions. We have to be aware of the two alternative visions and we have to choose to work for Zion. That is the spiritual message which lies behind the environmental problems of our era.

We are not creating an eco-topia like some Georgetown Guyana romantic kingdom. It is much more subtle and involves little gems of work, little cuts to a diamond each day in the visions we help to create, in the relationships which we build with spiritual values of trust and peace and caring rather than in using people for competitive advantage, or in the actual physical expressions of those values in technology and bricks and mortar.

We can build a city which is based around greed and selfishness where people can escape the demands of the world and of their neighbours. Such a city will also be polluting and will devour resources without restraint, or we can build in other values - it is not just an individual exercise but it starts there.

So this vision suggests that the spiritual dimension of any physical reality in which we work is constantly being added to the city. It gets built in. Our work has real meaning and purpose - we are agents of hope. This message of hope must now ring out loud and clear. In the '1980's the doom awareness was there - but where was the hope awareness?

People need hope otherwise the sense of looming disaster can lead to:

- **desperation** (even suicide where statistics reveal astonishing increases amongst younger people who see little future for themselves), and

- **apathy** (if there is no hope then why not just look after oneself).

We need hope but we also need to understand what that means.

Theology of Hope

Hope is a fundamentally religious word. It is used countlessly in the scriptures of all religions. It has been developed into a rich theology by Jurgen Moltmann in his book "The Theology of Hope" where he says that hope is not a feeling it is a choice.[15] You must choose to hope. The question is not "do you feel hopeful" but "do you hope".

Now it is my firm belief that this kind of hope has been working away in the middle of all those crises that I outlined before. It is almost never a media event however and so many people are unaware of these symbols of life.

Let me outline some of the signs of hope in the human environment which I have seen as emerging and which need to be told in the 1990's. What are the signs which can help us to give direction in our world?

The Sustainable Development Debate

One of the most debilitating elements of the environmental debate has been the inability to draw together ecological and economic realities. One of the most common responses to the environmental problems I outlined is to say stop all developments, stop any further economic growth as it is the economic system which is fuelling all of our problems. GNP is renamed Gross National Pollution, all industry is seen as exploitive, dangerous and polluting, adding greenhouse gases and consuming precious water supplies, all in the name of progress but in fact just adding more affluence to those who already have enough. With economics painted black there has been no context for providing solutions since economic institutions are a fundamental part of the social framework.

However this debate has neglected the work of many good econ-

15 J. Moltmann, Theology of Hope, London: SCM Press, 1967.

omists who have shown how the economic system can be adjusted to help solve ecological problems. And most recently we have been given a book which rings out with hope by drawing together the ecological and the economic.

This book is the UN Bruntland Commission's Report on Environment and Development published in 1987 called "Our Common Future".[16] This book and in particular its short summary should be made available to all students of the future world and its possibilities. Mrs Bruntland, the Prime Minister of Norway, and her Commission have done a remarkable job at showing what the problems are and at the same time showing that solutions are possible.

They have drawn the quest together into a phrase - "sustainable development". This means that the world needs development but it must be of a kind that does not compromise the ability of future generations to meet their needs.

This concept has three major principles:

1. The world must move towards sustainable yields from forests, from the soil and from the ocean.

2. The world should grow economically so that the poverty in the third world can be overcome as this is the main reason why the rainforests are being cleared, soils overgrazed and fisheries depleted and why populations continue to grow.

3. The world must develop industries that use energy, water and materials more efficiently and have the highest environmental standards.

These principles are then applied to show how the social and political framework needs to incorporate these changes.

---

16 World Commission on Environment and Development, Our Common Future, Oxford: Oxford University Press, 1987.

This is an important book because it sets out the context for hope, the framework for a future where these problems can be resolved. In Canada and New Zealand, the concept has been enthusiastically adopted by government, industry, unions and environmentalists. A Canadian 'accord' has been signed setting out the principles for future development. In Australia, the meaning and relevance of sustainable development is being set out by the Commission for the Future and the Prime Minister's Ecologically Sustainable Development Working Groups. In WA, we have been attempting to define our role in this global context. A recent publication, "Towards Sustainable Development in Western Australia",[17] which I helped to produce sets out the kinds of values and directions that I believe provide some hope for the future in Australia.

Global Stories of Hope

We need to know that things can be changed and improved - that is basic to hope - there are many such stories though we do not often hear them. For example:

- Forests are increasing at a greater rate than they are being cut in many parts of the world such as the USA.
- USA uses less oil today than it did in 1973 and California produces more power from the wind than all of WA's electricity generation.
- The great whales are returning to our oceans.
- Some countries like China, Korea, Cyprus have made enormous strides in rehabilitating degraded land.
- India and China both reversed appalling food import

17 State Government of Western Australia, Towards Sustainable Development in Western Australia, Perth: Ministry of Economic Development and Trade, 1989.

requirements in the 1970's to become self sufficient in less than a decade.

- Dramatic decreases in birth rates have been occurring in many Third World countries as a result of combined programmes in local grassroots economic development and family planning.

There are many success stories though the media tend to bias us towards the failures. Generally these stories can be found in UN publications though also in books like the Annual "State of the World" series by Worldwatch.[18]

My particular area of interest, cities, received a great boost recently with the formation of an International Urban Ecology Association - few people ever try to bring environment and city together. Suddenly it is beginning to happen and Australia is providing a lead.

Global Treaties

It is important to know that there are global processes at work today which are bringing together nations to solve global environmental problems e.g. the Law of the Sea, the International Treaty on Movement of Toxic Wastes and the Montreal Protocol on Ozone Depletion, have blazed a trail in international co-operation, unheard of until the 1980's. The Montreal agreement is an exceptional example of how a global problem can be addressed in a mere few years and how once accepted by a majority of nations it can facilitate an enormous change in consumption patterns and rapid changes in technology. By 1995 most CFC's will be gone.

There is a need for everyone to grasp a sense of "Our Common

---

18 Worldwatch Institute, State of the World - A Worldwatch Institute Report on Progress Towards a Sustainable Society, New York: W.W. Norton & Co., Annual.

Future" and see that there are processes and people and politicians who have achieved things in the implementation of this vision.

Local Case Studies

With the global scene as the backdrop the story of hope also has its examples in an Australian context. Each State is working on its own set of solutions to environmental problems. Each has stories to tell where solutions to environmental problems have been found.

In WA, I have tried to do this by producing a book called "Case Studies in Environmental Hope".[19] This book published in 1988 contains 18 case studies where individuals, groups, communities and government have faced up to an environmental problem and either solved it or set in motion the process of resolution. We have stories about:

- how the State's animal emblem, the rare and endangered numbat, has been successfully bred and released back into the wild;
- the remarkable restoration and rejuvenation of Fremantle;
- the saving of the Shannon, a WA symbol for conservation in the past decade every bit as important as Daintree and S.W. Tasmania;
- the scientific breakthroughs in controlling jarrah forest dieback;
- the process of stopping the destructive release of marginal land for farming;
- the bureaucratic and political breakthroughs in achieving strengthened environmental legislation; and

---

19 P. W. G. Newman, S. Neville and L. Duxbury, Case Studies in Environmental Hope, Perth: Environmental Protection Authority, 1988.

- the remarkable tree planting which is now occurring throughout WA.

A key feature of all these stories is that they involved a battle - they could have gone either way. It was not just awareness - there was commitment. Awareness is necessary though.

The involvement of schools in these processes is a great source of hope. The front cover of our book is a photo of school children landscaping a road verge in the country. Increasingly outdoor education is having some project attached to it which will enable students to have "hands on" experience and to begin to own some small symbol of environmental hope. Often this involves changes due to their actions. The great out-of-control world "out there" is not quite so remote and forbidding to children who have participated in such a project.

Any group can participate in similar projects and I would encourage everyone to do so. But especially I would urge everyone to consider taking on some real issues, some real battles. Global projects through international connections are probably the most important e.g. "Greencross" the new concept from Mikhael Gorbachaev, which is a voluntary organization of international people who will give up a year from their work to go to third world countries where people have impacted on the land rather than the reverse where the Red Cross are called for. There are many other such aid projects which need our support.

Let me finish by drawing together what this all means for us as ordinary people.

The Individual's Role

What is the role of the individual in a world which is "groaning as though in childbirth" as St Paul describes it.

In a word it is to **hope**.

Hope has two opposites which Moltmann develops - one he calls presumption, the other despair. [20]

Presumption says that **problems are quite simple to solve**. All you do is bring down a New Age, a new millennial vision. This generally involves violence - a new power is instituted through violent revolution of the left or right that purges away all those problems. This is the Prometheus solution - the one who stole fire from the gods. However the problems don't go away - Stalin's vision and Pol Pot's 14th century Buddhist vision killed a lot of people but did little to solve the problems they were trying to address. Baha'i's do not need to be told that these solutions do not work. We do not need to be a Prometheus but a servant.

The radical left and the radical right have little to offer in solving environmental problems as they do not hope, they presume.

Despair says there is nothing you can do. This is expressed succinctly by some existentialists e.g. by Camus whose motto was:

"thinking clearly and hoping no more".[21]

There is no shortage of existentialist environmentalists - the universities and coffee shops are full of them - lots of thinking and talk but no action. The environmental problems of today are a monument to the failure of this existentialism - defining your own reality is pretty limited when the ozone layer is burning your head!

As Moltmann says: "Neither in presumption nor in despair does there lie the power to renew life".

Hope sees that there is a battle - like a battle between two cities that can take our minds and hearts. We can retreat behind the city wall and shut out the nasty world, perhaps buy an occasional biodegradable cleaner but mostly blame "them" and get on with leading our own life. Or we can make a contribution, join the battle and cut a few gems in the streets of the other city?

Hope recognises the reality of our environmental problems and then takes small steps towards its solution. It is not wishful thinking, nor

---

20 J. Moltmann, Theology of Hope, London: SCM Press, 1967.
21 cited in Moltmann J., Theology of Hope,op. cit., 1967

is it a desperate leap into the future, it is a quiet clear vision that we move towards in small meaningful steps, generally with a group of others.

This is the final aspect I have always found to be of significance in solving environmental problems which helps us to reflect on the role of the individual. This is the role of the small group or small community. I am amazed at what a group of 10 to 20 can do to solve major problems locally and globally.

When Abraham tried to intercede with God about the problems of Sodom and Gomorrah he asked: "Would you reconsider destroying the cities if perchance there were 50 good people left?" "Yes" says God; then 40, 30, 20 and finally 10 is seen by God as the limit. He would not destroy a city if there were 10 good people. But Abraham can only find 2 or 3 - so the cities are destroyed.

Ten people is actually very few people to keep a city alive. I find in all the campaigns and "hope" activities in which I have been involved that very little is possible with 2, but with 10 like-minded committed people you can transform a problem into a real solution.

Jesus and most other spiritual leaders had a similar approach I have noted!

Finally may I suggest that you make for yourself a Hope List. This cannot be a list of the things you hope will happen somehow, somewhere else, but it is a list of things you choose to hope in.

You may already have a list you can draw up covering hopes which have moved a little, someway, a few steps closer to the world as it should be. Hopefully you will find more to work on from the other papers in this book.

Don't be presumptuous and don't be despairing, there is hope, but you must choose hope.

LOSS OF PRIME AGRICULTURAL LAND: AN EMERGING THREAT TO SUSTAINABLE DEVELOPMENT IN AUSTRALIA

David J. Chittleborough[1]

Introduction

In Australia it has been estimated that 400 ha of agricultural land is lost every day to urban sprawl, hobby farms and subdivisions, mining, parks, and recreation and holiday resorts.[2] Much of this loss is irreversible and is part of a world-wide phenomenon,[3] the import of which has only just recently been appreciated. The purpose of this paper is to highlight the problem and to suggest ways in which it may be overcome.

Loss of land by urbanization

Most of the irreversible loss of land takes place in the cities, especially the capitals, where land is paved, excavated or built upon. There are no reliable statistics for the rate of soil loss through urban growth, but an estimate is possible from the reports of the Australian Bureau of Statistics, which lists the reduction in the area of agricultural holdings in the major metropolitan areas. Table 1 shows the change in total area of rural holdings for Adelaide, Sydney, Brisbane and Perth. The figures are certainly an underestimate of the total area of land lost, because non-agricultural land will also have been developed for urban

---

1 Lecturer in Soil Science, Department of Soil Science, University of Adelaide, Waite Agricultural Research Institute, P.O. Box 1, Glen Osmond, SA, 5064.

2 Anon., "Why our farmland is disappearing", National Farmer, Thursday, 16 April, 1981, pp. 18-20.

3 J. K. Leslie and P. J. M. Johnston, Residential encroachment on rural lands, Australian Institute of Agricultural Science Occassional Publication Number 3, Melbourne, 1982.

purposes. This may account for the smaller loss of land in Sydney, because much development has taken place in areas which were never farmed, such as the north shore. At an average figure of 3,000 ha yr^{-1}, a minimum of 15,000 ha yr^{-1} are lost through urban expansion in the five mainland capital cities.

The worldwide loss of soil, even taking into account the projected loss to the end of the century, represents 2% of the cropland base. But the problem of soil loss by urbanization is much greater than is apparent from this statistic because cities are often located on the most fertile soil. In Canada, a study of changing land-use patterns showed that half of the farmland lost to urban expansion was coming from the best 5% of the farmland.[4]

Table 1: Reduction in area of rural holdings for four Australian cities.

City	Period	Average loss (ha yr^{-1})
Sydney	1948-1971	1,600
Adelaide	1956-1975	3,400
Perth	1967-1975	4,400
Brisbane	1970-1976	3,000

Source: Australian Bureau of Statistics.

Hallsworth has drawn attention to the fact that Australia's development proceeded on the blithe assumption that our land resources were enormous.[5] Prior to the 1930's little attention was given to their conservation. Exploitation was the keynote, but beginning in the 1930's conservation measures were applied to reduce damage from soil erosion caused by unwise agricultural and pastoral practices. But the transfer of

4 L. Brown, "Eroding the base civilization", J. Soil Water Conservation 36, 1981, pp. 255-60.

5 E. G. Hallsworth, "Where should we build our new cities? The national constraints." In, The Opportunity for Choice, ed. E.G. Hallsworth, Melbourne: CSIRO, 1978, pp. 1-7.

agricultural land to urban use was taken as a natural consequence of national growth.

Given that there is a population density of only 0.5 people ha^{-1}, it is not surprising that this view has prevailed. However, of the 769 million ha of land in Australia, only 77 million can be considered suitable for productive agriculture (Table 2). The productive land in Australia comprises 10% of the total area and of this almost 10% was urbanized as of 1973.

Another estimate of Australia's productive soils based on different criteria was derived from the "Atlas of Australian Resources" (Figure 1, Table 3).[6] On this basis only 3.5% of the Australian continent has a favourable combination of soil, climate and landform, but as can be seen from Figure 1 these areas lie in close proximity to the main centres of urbanization.

Under present conditions the land occupied by urban areas is probably < 0.5% of the world's land surface. The land used for urban settlements is only a small percentage of the land conceivably habitable. But within individual countries the proportion of land occupied by urban areas can be much higher. In parts of Europe it is as high as 10%. Most nations, including Australia, are not in immediate danger of running out of agricultural land. However, given the diminishing quality of the available additional land, Australia may encounter difficulty in providing enough of the right kind of land for particular purposes. Australia is a land extravagant nation with urban communities that have largely developed at low density (25 persons ha^{-1}) and which are highly dependent on the automobile.

Australia's position needs to be viewed in a world context insofar as rapid irreversible loss of land by urban encroachment is a worldwide phenomenon. In the United States nearly 400,000 ha of prime cropland

6 Division of National Mapping, Atlas of Australian Resources, 3rd Series, Vol. 1, Soils and Land Use, Canberra: Div. Nat. Mapping, 1980.

Figure 1. Distribution of soils with minor limitations for plant growth in relation to the climatic limits of dryland agriculture.

Table 2: Areas of land (x 10^6 ha) in various land use categories. Figures in brackets are proportions of the total land area in the particular category.

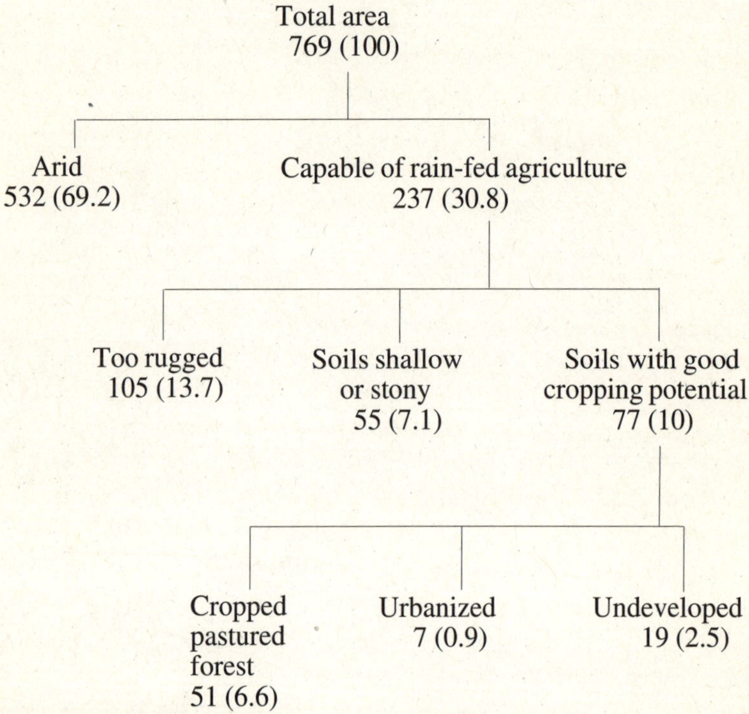

Source: After H.A. Nix, "Land use planning for commercial agriculture", **Proc. Symposium on Land Use Planning in North Queensland**, Australian Institute of Agricultural Science:: Brisbane, 1974.

is converted each year to non-farm uses. In Florida, which produces much of the world's citrus, all the prime farmland will be put to other uses by the year 2000 if the current trends continue. Over the same period, 16% of prime cropland will be lost in California, and in Virginia 24%. In the decade 1960-1970, West Germany lost 2.5% of its agricultural land, and France and the United Kingdom nearly 2% each. Third World figures are scarce but, at a conservative population increase of 1.35 billion people between 1980 and 2000 at 0.04 ha per person will mean a loss of 54 million ha.[7] If 40% of this is cropland, Third World cities will cover an area of 22 million ha, equal to the cultivated area of France.

Growth of Australian cities

Australia is one of the most highly urbanized countries in the world in terms of the proportion of people living in cities and towns: 40% of the population live in two cities, and 85% in settlements exceeding 1,000. In view of the fact that 70% of the Australian population live in 5 cities, the loss of land by urbanization is highly concentrated, more so perhaps than any other nation. Furthermore, the loss of land per head of population is the highest for any continent. This is because Australian cities are among the least densely populated in the world, or, put another way, Australian cities per head of population are among the largest cities in the world. The average gross density of 21 Asian cities with a million or more population is 100 persons ha^{-1}. The average of 17 such cities in Western Europe is 45, of 14 in Eastern Europe and the USSR 43, and of 25 cities over a million in the United States, 17. The average density of the 5 mainland capital cities in Australia is 15 persons ha^{-1}.

Australian's urban growth has its roots in the early settlement. As an island continent settled from abroad, the first white settlers were

7 L. Brown, "Eroding the base of civilization", J. Soil Water Conserv. 36, 1981, pp. 255-260.

Table 3: Proportion of land in Australia with few limitations for rain-fed agriculture (%).

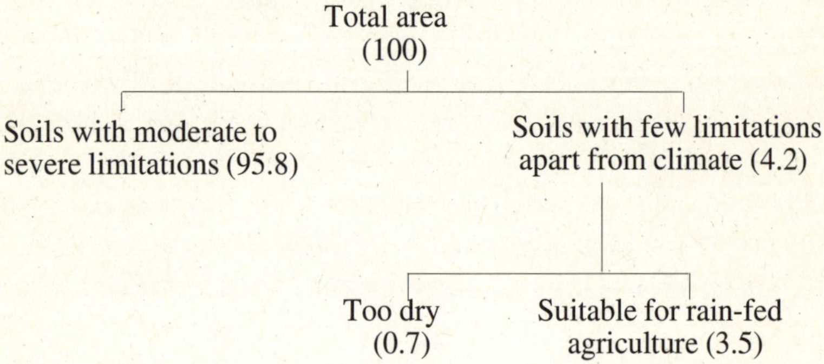

Source: Division of National Mapping (1980)

exclusively in coastal port towns. Penetration inland was hindered by an inhospitable environment, few navigable rivers and limited possibilities of an agrarian staple for immediate export. Further, the composition of the emigrant population was urban. As Bourne has pointed out,[8] Australia was urban before it was agrarian. The first settlers were townspeople who were not concerned with settling the interior of the continent and whose occupations were largely urban. The coastal port cities were the only locations of major settlement. Apart from Sydney, the capital cities and most of the major towns were located on or near good agricultural land with an assured water supply and rainfall.

Specific examples of encroachment

Cudgen-Duranbah, New South Wales: A recent study on the far north coast by the New South Wales Department of Agriculture

8 L. S. Browne, "Urban systems in Australia and Canada: Comparative notes and research questions", Aust. Geogr. Studies, 12, 1974, pp. 152-172.

highlights the problem of rapid urbanization.[9] The north coast region comprises a narrow coastal strip 120 km in width extending 450 km from below Taree in the south to the Queensland border in the north. The western boundary is formed by the escarpment of the Great Dividing Range. Although the region has a high rainfall there are extensive areas of soils of low fertility developed from sandstone and shale. These soils make up 77% of the area. The remaining 23% of the north coast land comprises fertile alluvial flood plains associated with eight easterly flowing rivers (15%) and krasnozem and related soils derived from volcanic rocks and sediments on low hills and plains (8%). The combination of productive soils and elevated coastal location provide for the longest growing season and the most diverse range of crops in New South Wales.

In the Cudgen-Duranbah area, where the Soil Conservation Service focused its study on the effects of urbanization, it found that the land was highly prized for retirement home sites and vacationer cottages and was under immediate and growing threat of urban subdivision (Figure 2). The areas of frost-free red basaltic soils at Cudgen-Duranbah and Terranora are important for the production of vegetable crops, particularly tomatoes, to supply the southern Australian markets from June to October. Collectively the productive soils of the north coast produce virtually all of the state's fresh tomatoes, beans, peas, sweet potatoes, zucchinis and cucumbers.

Queensland sugarcane land: Limited areas in Queensland are suitable for sugarcane production, but a recent study concluded that the soils on which sugarcane is grown "are subject to significant residential pressures with serious implications to long-term future of the industry in some production areas".[10] Sugarcane grows best on frost-free, high-

---

9 R. J. Smith, Cudgen-Duranbah agricultural land suitability study, Sydney: NSW Dept Agriculture, 1982.

10 Australian Institute of Agricultural Science, Residential encroachment on rural lands, Report of land use sub-committee, Qld Branch Bulletin

rainfall land or soils capable of irrigation. Such land occurs in specific locations along the Queensland coast, notably Mossman-Ingham, Ayr, Mackay, Bundaberg and Nambour (Figure 3). A study by the Bureau of Sugar Experiment Stations has indicated that the actual loss of assigned lands to urban development in the 18 shires of Queensland where sugar is grown amounted to 2,250 ha between 1972 and 1976. A further loss of 1,600 ha of land was predicted in the 5 years from 1976 to 1982. The loss has occurred mainly in urban areas such as Cairns and Mackay which are surrounded by canelands. While the losses represent only 1.2% of Queensland's sugar lands, the urban pressures are increasing dramatically. Costs will escalate as a result of production being forced on to less suitable land. For example, in one location production has been relocated on to reclaimed saline land.

Proposals to overcome urban sprawl and the consequences for loss of soil

There have been three proposals to overcome the urban sprawl and all involve government intervention in the urbanization process: (1) the establishment of new centres, (2) a stimulation of urban development in rural townships - the so-called growth-centre concept, and (3) urban consolidation.

New Cities: Hallsworth suggests that there are two areas which are favourable for the establishment of new cities:[11] the coast of the Great Australian Bight in South Australia and the Queensland coast 250 km north of Brisbane. Minimal damage would take place to the soil resources because both areas have stony, infertile soils. A choice of these areas would take the pressure off southeastern Australia. Such development

no. 197, 1976.

11 E. G. Hallsworth, Where should we build our new cities? The national constraints. In, The opportunity for choice, ed. E. G. Hallsworth, CSIRO: Melbourne, 1978, pp. 1-7.

Figure 2: Cudgen-Duranbah area
Distribution of krasnozems at Kingscliff on the far north coast of New South Wales.

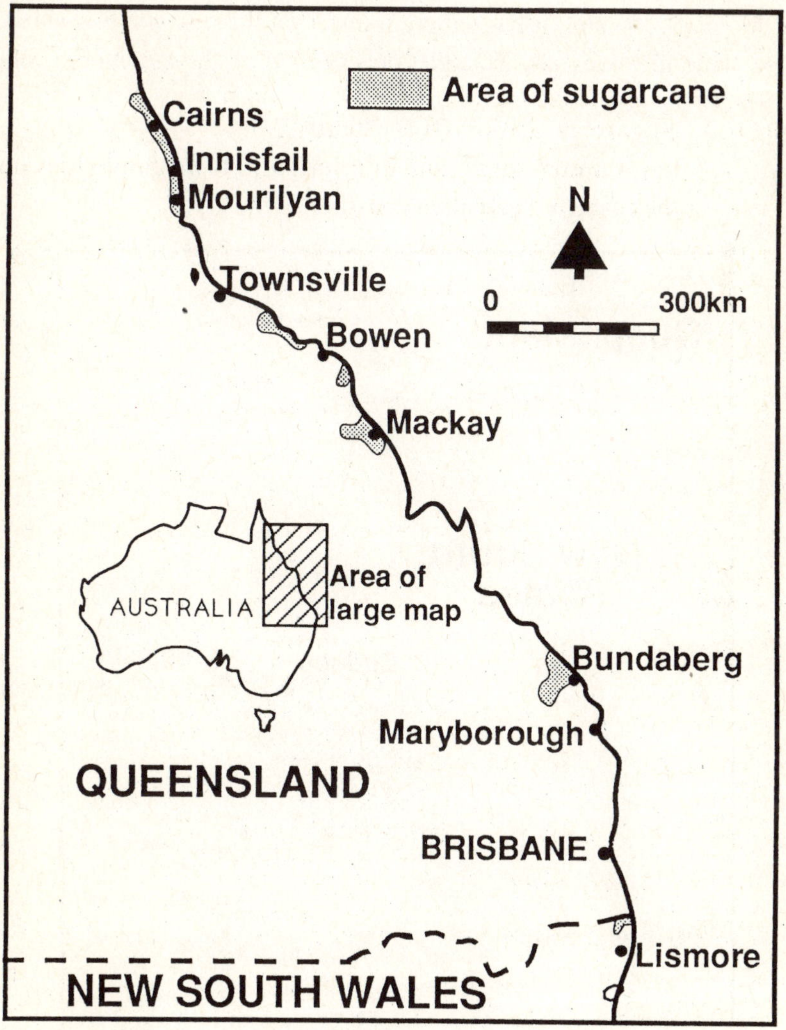

would require federal government intervention and has the disadvantage of requiring a large government and private investment.

Growth Centres: The early 1970's saw the emergence of the growth-centre concept as a means of fostering decentralization. In terms of its impact on soil, it has simply transferred the problem elsewhere. The same lifestyles, for example, dependence on the automobile and

Figure 3: Sugarcane area of Queensland
Land under sugarcane in relation to towns and cities on the eastern Australian coast.

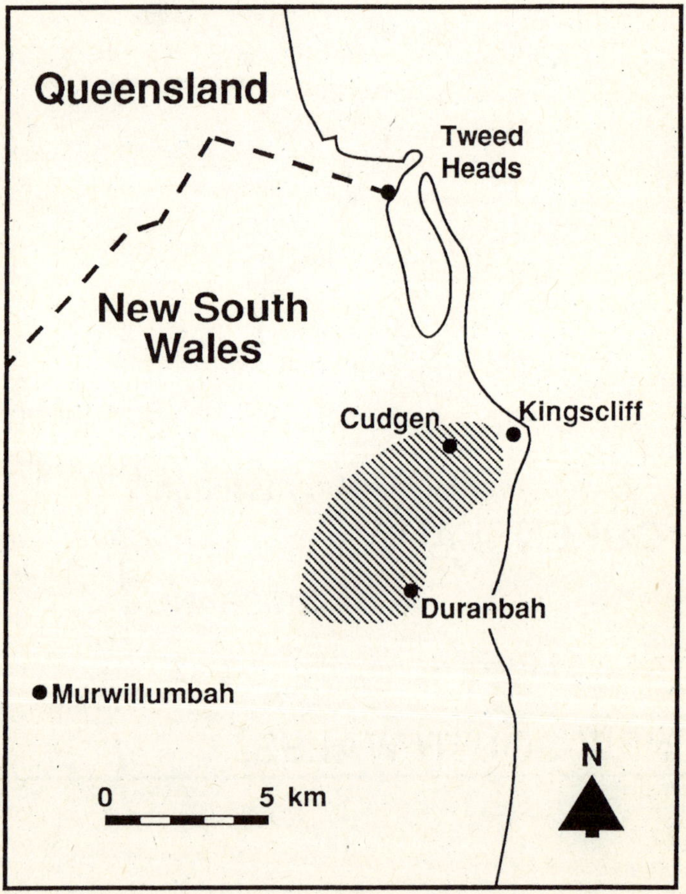

large residential blocks, are in evidence. We are simply paving over another part of Australia.

Urban Consolidation: In contrast to the prevalence of double-storey and multiple-storey dwellings in many other countries, Australian cities have extensive areas of single-storey dwellings on large blocks of land. For example, the density of Perth's most populous suburb (Subiaco) is 24 persons ha^{-1} and the average for the whole city is 12.2. For cities in Britain the average is 33.8. The current density of new development on the urban fringes in Australia is about 15 persons ha^{-1}; hence the urban sprawl. (The Greater London Council aims for a net residential density of 336 persons ha^{-1}.)

A change in urban form is the most environmentally sound option, but the concept is meeting with considerable resistance from inner city dwellers who are established in their single storey house on a 1/4 acre block. There is also resistance from Australians who are conditioned to ownership of a home with plenty of surrounding land. The sociological problems arising from poorly developed multi-storey tenement blocks in some countries in the 1960's and 1970's has also led to concern about urban consolidation. These attitudes can best be changed by governments (federal, state and local) demonstrating the benefits to urban residents; better services, greater economy and (with the aid of some imaginative architecture), more attractive environments at less cost. The diminished pressure on surrounding cropland would be a fortuitous by-product. However, the saving of prime agricultural soils must be moved to the centre stage of urban planning policy for environmental and economic reasons.

A change of attitude

Most of the environmental problems relating to unwise land use have been laid at the doorstep of the farmer. Salinization, wind and water erosion from overcropping, overgrazing or overclearing, waterlogging, acidification, even eutrophication of streams and watercourses, have been attributed to farmer ignorance, greed or ineptness. The urbanite has been a bystander. But this is not so with urban encroachment. Urban dwellers are not divorced from this problem; they are an integral part of it and must be involved in its solution. A change of attitude is required, by governments, industries and individuals, in relation to the world's soil resources.

In the summary of "Our Common Future", it is stated that "we must collectively stop using up the Earth's ecological capital and begin to draw on the interest we can get from the sustainable husbandry of its resources. In order to accomplish this transformation, governments need to do two things. First, they need to make the concept of 'sustainable development' central to all planning and activities. Secondly, the community of nations needs to evolve a new, more equitable, international economic structure that begins to narrow the gap between developed and developing countries. The need to integrate economic and ecological considerations into development planning will require nothing less than a 'renaissance' in economic decision making".[12]

The future of Australia's (indeed the world's) prime agricultural soils depends on our commitment to this approach.

Conclusion

In absolute terms there is enough agricultural land in Australia to support the population in the foreseeable future. Given the diminishing quality of the available additional land, however, Australia may face serious difficulty in the future in providing, firstly, enough of the right kind of land for particular purposes and, secondly and perhaps more importantly, land of quality which will avoid the rising economic and environmental costs of agricultural production. A change in urban planning policy is required in Australia in which the protection of prime agricultural land is a central policy.

---

12 World Commission on Environment and Development, Our Common Future, New York: United Nations, 1987.

ENVIRONMENTAL DAMAGE: RESHAPING INTERNATIONAL RELATIONS

Keith D. Suter[1]

Introduction

This presentation is about the changing nature of how nations deal with one another ("international relations"). The current environmental crisis is obliging governments to rethink international relations. The rethinking began before the current environmental crisis - and it will continue, with or without the crisis. But the crisis has been an added inducement to do that rethinking.

This presentation covers four matters. First, despite the now fashionable references to "common crisis" and "common future", it is necessary to remind ourselves that there is far from widespread agreement on the need for a specific, directed, purposeful change in international relations. No one denies that changes are underway. The problem is one of who (or what) should guide the pace and direction of the changes. In short, the assumption underpinning this conference may not be shared by all governments.

Second, considerations of changing international relations need to be done in the context of the Brundtland Report: "Our Common Future".[2] Consequently, this presentation summarizes the main points of the Report's Chapter 12 dealing with institutional and legal change.

Reference has already been made to the proposed changes to international relations that were already being debated before the current

---

1 Executive Director, National Goals and Directions Movement, G.P.O. Box 4878, Sydney, NSW.

2 World Commission on Environment and Development, Our Common Future, Oxford: Oxford University Press, 1987.

environmental crisis. The third task of this presentation is to summarize the main proposals, arising from non-governmental organizations and private individuals, for reviving the United Nations (UN) system.

Finally, there is a case study of the pioneering work done by Col Howard and the late Rev Harriet Kurtz. For those of us involved in the earlier environmental crisis (late 1960's/early 1970's), much of the current concern has a familiar ring. A crisis is identified; some governmental action is introduced; public apprehension recedes; attention is then focussed on some other crisis. The Kurtzes, by contrast, have plugged away relentlessly on some basic ideas; public interest has risen and fallen; but the Kurtzes have just kept ploughing on - safe in the knowledge that eventually their ideas would be accepted.

Are We Agreed on "Common"?

"Common" is now very common! The 1980's saw a string of international independent commissions of enquiry talking about, for example, "common crisis" and "common security"[3]. The most well known was Mrs Brundtland's "Our Common Future".[4] The use of the word "common" was not accidental. It reflected the prevailing philosophy of the need for a common response to common problems. But is that philosophy accepted by everyone? It is necessary, then, to begin this presentation with a note of caution. The assumption underlying this conference is not necessarily shared by all governments - or, perhaps more accurately, all governments have, to varying extents, to reconcile two contrasting philosophies: individualism versus common action.

---

3 The Brandt Commission, Common Crisis North-South: Co-operation for World Recovery, London: Pan, 1983.
Independent Commission on Disarmament and Security Issues, Common Security: A Programme for Disarmament, London: Pan, 1982

4 World Commission on Environment and Development, ibid.

Individualism

According to this viewpoint, the human being is the basic unit of account. Society should operate with the minimum possible rules and responsibilities. It is a matter of the "survival of the fittest".This viewpoint may be illustrated by the view that governments should not meddle with the economy (where they invariably do more harm than good). The economy is best left to the market system, where the organising power of self-interest will maximize self-gain for all. This is now called by some "New Right" thinking.

A well known advocate of this viewpoint is Milton Friedman, who has become the best-known modern champion of laissez-faire economics, and who sees himself in the great tradition of the free marketeers from Adam Smith onwards.[5] The Friedman message is one of: small government is beautiful; low taxes are possible; the pursuit of personal economic self-interest, although it may not be the highest of ideals, produces the best and fairest results. Capitalism can give rise to dramatic economic growth, such as Hong Kong, Japan, West Germany and Israel.

This viewpoint has now captured the minds of most Western Governments. President Reagan and Mrs Thatcher were the two most well-known political leaders who have tried to implement this viewpoint. Even Labor parties and Labor governments have moved in the New Right's direction. For example, the Chifley Labor Government in the late 1940's, tried to nationalize Australian banks - the Hawke Labor Government in the mid 1980's moved to the other end of the political spectrum and deregulated the banking system!

A New International Economic Order

Not everyone has benefitted from this brand of economics.[6]

---

5 Milton and Rose Friedman, Free to Choose: A personal Statement, Macmillan: Melbourne, 1981.

6 Sibusiso Bengu, Mirror or Model? The Church in an Unjust World,

Africa's poverty is a good example. There are three kinds of African poverty: emergency poverty brought about by a catastrophe such as drought or earthquake; intrinsic poverty caused by an absence of natural resources; and "induced poverty" which human policy has imposed upon the weak. Most poverty in the world is of this last type. Ironically, that poverty is often induced by the economic policies which were hailed at the beginning of the UN's First Development Decade 30 years ago as the solution to Third World poverty. These policies emphasized the need to increase foreign exchange earnings by building up exports of raw materials and cash crops, and the need for industrialization with advanced technology, building up the modern sector and big cities. Greater prosperity for all, it was said, would "trickle down" from greater prosperity at the top. That advice overlooked the need to balance capitalist expansion with progressive taxation, social legislation, and strong labour unions to aid the redistribution of income. Otherwise "money makes money" - and so money flows most readily to those who already have it.

The colonial era left a world economic order which makes it difficult for poor countries to develop. Colonial powers turned self-sufficient African economies into monocultures producing one or two commodities for export to the West. African economies are typified by lack of local industry and dependence on one or two primary commodities. Zambia's exports are 90% copper, Mauritius 90% sugar and Gambia 85% groundnuts. Prices for these raw materials fluctuate wildly making it hard to plan ahead.

World commodity prices have declined because of overproduction and yet the West still insists that less developed countries should expand export crops in order to develop. The developed world still pressures less developed countries to open up their markets to the manufactured goods from the West yet they themselves put up import

---

Lutheran World Ministries: New York, 1984.

controls, tariffs and quotas to deter less developed countries from processing their raw materials before export. The worsening terms of trade faced by less developed countries - an increasing tendency towards having to sell raw materials cheap and buy manufactured goods dear - can be seen as a transfer of income from the poor countries to the rich. This is a sort of reverse social welfare, with the Third World helping the First World (including Australia)! Interest payments, debt repayments, private capital sent abroad and repatriated profits all added up, exceed capital inflows in the form of aid, private loans and investment.

Private investment accounts for around two thirds of the capital flow to the poor world. But privately-financed development is not the type that would happen if the welfare of the poor were of concern. There is little profit to be made from schools, hospitals, roads and clean water supplies.

An additional problem arises out of the lack of capital accumulation - each Third World nation should re-invest its own profits so as to expand its economic activities. But local capital accumulation very rarely occurs. When an economy is dominated by transnational corporations and allows its policies to be dictated by the requirements of transnational corporations, it can never indulge in this process. It must open its economy to allow the continuous repatriation of profits. This puts a brake on the process of capital accumulation. Once an economy allows itself to be dominated by transnational corporations, it automatically allows its technological development to be dependent on these agents. It will get the kind of technology that has little relevance to its real requirements so as to fit the requirements of the transnational corporations.

In June 1974 the UN General Assembly called for the creation of a New International Economic Order (NIEO). NIEO is, in essence, an affirmation of the need to correct inadequacies in the present economic order by creating a new pattern of global relationships based upon equity, common interest and cooperation among all nations. Third World nations have called attention to the debilitating impact of their dependence upon raw material exports and, in particular, to the fact that their overall earnings cannot keep pace with the rising costs of imported food and manufactured goods. NIEO lays out a wide-ranging set of measures designed to help establish a more just global economic order.

Seventeen years later, the NIEO project is virtually dead. Indeed, the phrase itself hardly appears in UN documents. First World governments and transnational corporations have won the day - let the market (and not human intervention) determine the pace and direction of international economic relations.

This is a poor omen for trying to create a new system for international environmental relations!

Gorbachev Discovers the Market

Just to add to the gloom, it is worth recalling that a most recent convert to this view point is Mr Gorbachev. Mr Gorbachev maintained that radical reform of the Soviet economy was unavoidable if the Soviet Union was to enter the 21st century as a world economic and political superpower.

Self-sufficiency for each individual Soviet economic operation holds the key to the system of "Perestroika". Instead of massive subsidies being handed out almost regardless of production efficiency, firms will now be strictly confined to working within the limits of direct returns from production. All purchases of raw material purchases, plant modernisation costs and labour charges will have to be balanced against each factory or farm's profits. Loss-making enterprises will risk being closed.

Wage policy is also to be significantly changed. Workers are to be paid according to how much they produce individually and its quality. Employees who are either unable or unwilling to work hard are being faced with a cutback in wages or the sack. The management and the workforce of each enterprise will be allowed to distribute funds in any way they choose. It is in the interests of both parties to sacrifice jobs without an accompanying fall in production, so that per capita wages rise. Consequently unemployment will increase.

Mr Gorbachev was a long way from speaking openly in favour of Adam Smith and Milton Friedman. But in all the euphoria of 1989's "Fall of the [Berlin] Wall" etc, it is worth noting that this means, in essence, a rediscovery of self-interest, "rational" economics and placing the individual ahead of the community.

I sound the note of warning because we must not become too overwhelmed by "Gorby-mania". Certainly, Mr Gorbachev was a welcome advance on all his appalling predecessors. Additionally he has foreshadowed (albeit in a vague way) that the USSR will play a more constructive role in the UN. But the logic of his viewpoint - if he is to be consistent - means permitting the market to set the pace and direction of change in international relations.

The Common Interest

An alternative viewpoint of the individualism of the New Right is to emphasize the common interest. For individualism is ultimately self-defeating.

This point may be illustrated by the Prisoner's Dilemma. Two people commit a crime. They are arrested by a police officer acting on a hunch - rather than having any firm evidence. They are placed in separate cells. The officer interviews each one in the hope that one or both will confess. The officer offers each one a deal: if one confesses, then he will get two years (as a Crown witness) but the other will get the full 10 years. They cannot confer. If both keep quiet, the officer will have to release them because of the lack of evidence.

The Dilemma for each Prisoner can be set out as follows:

Options	A	B
	Number of Years in Gaol	
If both confess	10	10
If A confesses	2	10
If B confesses	10	2
If both keep quiet	0	0

This is a vivid example of the failure of self-interest to achieve what it seeks. From the point of view of self-interest, each criminal will confess in the hope that the other will not - with the expectation that he will get only two years in gaol, while his colleague gets 10 years. Thus both confess - and both get 10 years. On the other hand, if both assume that the other is like himself - trustworthy of a sort - both leave gaol immediately!

The Prisoner's Dilemma is seen in environmental terms each day. A nation does little to curb its own pollution both because such a curb could increase the cost of production (and ultimately make the product more expensive) and, anyway, the pollution is carried offshore by the wind or sea. If that example is followed by other nations, then all nations will eventually become polluted by transborder pollution. If every nation seeks to maximize self-gain, then it will be to its own eventual disadvantage.

Tension between the individual viewpoint versus the common interest viewpoint goes in cycles. At present, the former is the more fashionable. Immediately after the Depression and World War II, the latter was the more fashionable. Those tragedies forced upon governments the need for international co-operation (the major manifestation of which was the creation of the UN). As the former viewpoint returned to vogue - when people forgot the lessons of the Depression and World War II - so international co-operation was overtaken by international competition and the UN's evolution was halted. It is no coincidence, then, the US and UK leaders who have been most hostile to the UN since 1945 were President Reagan and Mrs Thatcher.

Perhaps the emerging environmental crisis will oblige conservative leaders to pay more attention to the work done by non-governmental organizations and private individuals to reforming international relations. The rest of this presentation is about three such examples of that work.

Our Common Future

The final chapter of "Our Common Future" (Chapter 12) is

entitled "Towards Common Action: Proposals for Institutional and legal change".[7] In the context of reshaping international relations in the interests of protecting the international environment, this chapter is a useful starting point.The report makes it clear that the onus for action lies within no one particular group of nations. All nations have a role to play in securing peace, in changing trends and in creating a fair international economic system.

The chapter sets out what are essentially conceptual guidelines for institutions at the national level. Suggestions are also made for reforming international institutions.

One proposal is that governments should consider developing a "foreign policy for the environment". A nation's foreign policy needs to reflect the fact that its policies have a growing impact on the environmental resource base of other nations and the common areas (such as the seas), just as the policies of other nations have an impact on its own. In this regard, I have proposed that each nation establish a Ministry for Peace so as to bring together considerations of foreign policy, foreign aid, disarmament and environment.[8]

The report recommends an expansion of the work of the UN Environment Programme. There should be greater attention to global environmental assessment and reporting.

Scientific groups and non-governmental organisations should also play a much greater role in the international protection of the environment. There should also be improved links with industry.

A universal declaration and a convention on environmental protection and sustainable development are also proposed. A draft of the

---

7 World Commission on Environment and Development, *Our Common Future,* Oxford University Press: Oxford, 1987.

8 Keith D. Suter, The Australian Campaign for a Ministry of Peace, UN Association, 1984.

principles to be followed is published as Annexe 1 to "Our Common Future".[9]

The international financial institutions such as the International Monetary Fund and World Bank should give greater attention to the need to protect the environment when making loans. Bilateral aid agencies should also bear in mind the importance of protecting the environment when making funds available for projects.

An international report such as "Our Common Future" can at most hope only to identify areas of concern and to lay down general principles. It is not possible to be too specific. However the report is at least useful for raising some interesting challenges which need to be followed up.

Reforming the UN

Reforming the UN is now an item on the international political agenda. The UN has certainly made great progress but there is still much to be done. Various ideas are circulating for reforming the UN. There are few proposals for its total abolition - even its critics prefer to improve it rather than abolish it.

The immediate need for reform has been forced upon the UN through its current financial crisis - largely, but not solely, initiated by the United States. The Soviet Union, meanwhile, has for the first time since 1945 made proposals (albeit rather vague) for enhancing the UN's effectiveness. The main consistent source of proposals has come from non-governmental organisations (NGO's) and private individuals. Incidentally, it is worth recalling that NGO's were among the architects of both the League of Nations and the UN. I sometimes feel that NGO's take the UN more seriously than do many governments!

Non-governmental proposals vary considerably in their ambition. Some would like to see changes in practice (as distinct from changes to

---

9 World Commision on Environment and Development, ibid.

the UN Charter), while others want amendments to the UN Charter. Some would prefer a fresh attempt outside the UN.

All that can be done in this presentation is to illustrate some of the breadth and imagination of the proposals, with contact addresses for people wishing to obtain more information.

Reforming the Security Council

The Canadian Branch of the World Federalists has proposed some changes to the practices followed on the UN Security Council.[10] The Security Council is supposed to deal with threats to international peace and security but it is often hindered by the veto powers of the Big Five permanent members (US, USSR, UK, France and China). It also has 10 elected member-nations.

The Branch has proposed that there be an agreement among the permanent members to refrain voluntarily from using their veto in the following cases: ascertaining the facts; dispatching observers (with host country acceptance); functions of the Secretary-General in dispute settlement; where a permanent member is party to a dispute; adoption of resolutions calling for ceasefire; separation of armed forces; withdrawal behind borders; admission of new members and the establishment of subsidiary organs.

Environmental matters have rarely appeared on the Security Council's agenda since the Security Council only deals with immediate threats to international peace and security. But if the environmental crisis is going to become as great as currently predicted, then the Security Council will need to find ways of dealing with this crisis.

---

10 World Federalists of Canada, 145 Spruce Street, Suite 207, Ottawa, Ontario K1R6P1.

Reforming the General Assembly

The General Assembly contains all the UN member-nations. The total membership has now risen to 160 (with Namibia's recent independence). In the General Assembly, small nations have a voting strength disproportionate to their populations and political-economic-military strength; Botswana and Vanuatu have the same vote as the US and the USSR. Moreover, Assembly decisions are only recommendations which can (and very often are) flouted with impunity. By contrast, Security Council decisions are binding on all UN member-nations. But in the Security Council, no action can be taken without the approval of all five permanent members; the veto has crippled the Security Council.

The Binding Triad proposal is being advocated by the New York-based Centre for War/Peace Studies, headed by veteran journalist Richard Hudson.[11] The proposal consists of two elements. First, the Binding Triad would change the voting system in the General Assembly from its present one nation/one vote basis to a weighted voting arrangement based on three factors: one nation, one vote; population; and contributions to the regular UN budget. In order for a resolution to pass, it would have to win two-thirds majorities of these present and voting on all three legs. On the first leg, with 160 members in the Assembly, and all members voting, 106 would have to vote in favour in order for a resolution to pass. It is on this leg that the many small nations of the world have their greatest influence. On the second leg, based on population, a resolution must garner support from nations representing two-thirds of the population of those present and voting if it is to be adopted. Thus the most populous nations such as China, India, the Soviet Union and the United States could have the most effect on the outcome of a vote on the second leg. The third leg is based on contributions to the regular UN budget, which are roughly apportioned according to gross national product (GNP). In turn GNP is one approximate measure of the political,

11 Centre for War/Peace Studies, 218 East 18th Street, New York, NY

economic and military power of nations. On the third leg, therefore, the big industrialised nations would have the lion's share of influence. For a resolution to pass, it must win two-thirds majorities on all three legs of the Triad. When it does, it will have the support of most nations of the world, most of the people of the world, and most of the political, economic and military power of the world.

Second, the powers of the General Assembly would be increased under the Binding Triad so that in most cases its resolutions would be binding, and not just recommendations as at present. However, the Assembly would not be permitted 'to intervene in matters which are essentially within the jurisdiction of any state', as laid down in the UN Charter. If the jurisdiction were in doubt, the issue would be referred to the World Court, and if the Court ruled that the question was essentially domestic, the Assembly could not act. Furthermore, the Assembly would be prohibited from employing military forces to carry out its resolutions; this would remain the prerogative of the Security Council (as at present). However, under the Binding Triad system the Assembly could provide peacekeeping forces to carry out a mandate. Binding resolutions could be passed not only on peace and security questions, but also on economic and environmental problems.

The Binding Triad would be introduced into the international system by amending Articles 13 and 18 of the UN Charter. These provide for the voting system and powers of the General Assembly. Amendments to the UN Charter require acceptance by two-thirds of the member nations, including all five permanent members of the Security Council, and so the incorporation of the Binding Triad will not be easy. However, Hudson believes that a trade-off may be possible. The small nations do not like the Triad half of the Binding Triad because it means giving up their automatic majority in the General Assembly, but they would like to see resolutions become binding. It is precisely the opposite with the Security Council permanent members - they would like to have more power in the Assembly, but are reluctant to make its decisions binding. Both groups must be persuaded that in their mutual interest they must build a sensible global system of decision-making.

Representation of the People

The UN Charter begins with a ringing statement about 'we the peoples...' but then references to the 'people' disappear in favour of the UN being an association of nations (and governments). How, then, to improve the representation of the people at the UN? One proposal is for a Second Assembly. Another proposal throws the representation net still further and suggests two Chambers additional to the UN General Assembly: one for business corporations (which have little direct involvement in UN affairs) and one for the general public.

The campaign for a UN Second Assembly originated in the written statement of the UK Medical Association for Prevention of War (MAPW) to the UN Special Session on Disarmament in June 1982.[12] A Network of organisations and interested individuals to campaign for a UN Second Assembly was initiated by MAPW in March 1983. A UN Second Assembly would be an assembly composed of non-governmental persons. It would be linked to the General Assembly under Article 22 of the Charter and would operate under a Constitution granted by the General Assembly. Article 22 reads: 'The General Assembly may establish such subsidiary organs as it deems necessary for the performance of its functions'. The objective of the proposal is for the peoples of the world to have a representation in the UN as global inhabitants. This would supplement the existing General Assembly representation as national citizens. The bicameral representation of everyone would be somewhat analogous to the US Congress and the Australian Parliament, where the whole population is represented in both the Senate and the House of Representatives: the Senate represents the states, with each

12 International Network for a UN Second Assembly, 308 Cricklewood Lane, London NW22PX.

state having the same voting power (as does each member-nation in the General Assembly), while in the larger House of Representatives the inhabitants of the whole country are represented proportionate to population.

The three main functions of the Second Assembly would be to establish a dialogue between representatives of the governments collectively and the world's non-governmental movements for disarmament, development, the environment, human rights, prevention of armed conflicts, and prevention of the threat of nuclear war. Second, to support and stimulate organisations and institutions active in these fields (within limits allowed by the Assembly's Constitution). Third, to foster at non-governmental levels international understanding (the Second Assembly would not take sides in disputes between UN member-nations).

The selection of the representatives would be carried out within guidelines laid down in the Constitution of the UN Second Assembly. This should allow for variations in method in different nations. The number of seats allotted to each nation would be related to population size (that is, analogous representation to that in the US and Australian Houses of Representatives).

The number could be proportional, for example 1 for up to 1 million, 10 for 100 million, 32 for 1000 million. This method would give about 550 representatives. The members, being representative of the peoples of the world as global inhabitants, would be grouped into regions, not nations or political parties.

Marc Nerfin, a former UN civil servant and now director of the Geneva-based International Foundation for Development Alternatives, suggested in 1985 a total of three chambers for the UN General Assembly.[13] The 'Prince' Chamber would represent the governments of the

13 Marc Nerfin, International Foundation for Development Alternatives, 2 Place du Marche, Nyon, CH-1260, Switzerland.

nations. The 'Merchant' Chamber would represent the economic powers, be they transnational, national or local, belonging to the private or national sectors. The 'Citizens' Chamber, where there should be as many women as men, would, through some mechanism ensuring adequate representation, speak for the people and their associations. At the very least, this would make it possible for citizens to hold Princes and Merchants accountable for the consequences of the exercise of their power.

In early 1988 Nerfin set out the difference between his proposal and the previous one. He said that the International Network for a UN Second Assembly acts as if its proposal were implemented (with all the self-imposed limitations, like suggesting that such a Second Assembly be a 'subsidiary body' of the General Assembly or that member nations decide on representation methods). By contrast his idea of a 'Citizen' Chamber, was not a blueprint but rather a way to show the limits of the present UN system. It was, at most, a signpost to suggest the direction in which steps should be taken to give the global civil society some influence on governance of human affairs.

CAMDUN

CAMDUN took place in October 1990 in New York City: the first of four international "Conferences on a More Democratic United Nations". The second will be held in Europe and the third and fourth in India and Australia, respectively. Proposed in 1989 by INFUSA, the International Network for a United Nations' Second Assembly, as "an open forum for discussion of all relevant proposals on the strengthening of the UN system by democratic input from non-governmental organizations (NGO's), the public, and parliamentarians", the Conference is co-sponsored by a wide range of NGO's concerned about UN reform (but not, apparently, the Baha'is!)

According to one of the Camdun convenors, the timeliness of the theme, "Building a More Democratic United Nations", was implied in the address of Dr Noel Brown, Director, North American headquarters of UN Environment Programme, before a joint meeting of the UN/NGO Committees on Youth and the Aged, February 8th 1990. After reviewing the significant agreements of the 1989 Session of the General Assembly on environmental issues, he foresaw that the World Conference on Environment and Development in Brazil for two weeks in June, 1992, (the 20th anniversary of the Stockholm Conference on the Environment) would "promote the development of international environmental law." He declared that the Brazil Conference will evaluate the "adequacy of prevailing international institutions" and concluded: "the global agenda is no longer exclusively a government-driven agenda - and will increasingly become people-driven...".[14]

That last phrase, incidentally, is a nice play on words. The New Right likes to refer to the international (and national) economy as "market-driven". Now, it seems, we are to have "people-driven" policies!

CAMDUN will be only one step in what has already been a long process - and what promises to extend far out into the future: international institutional reform. It also serves as yet another reminder of a serious area of neglect. Non-governmental organizations campaign on issues and not on institutions. In other words, the peace movement devoted a great deal of energy to see out the dangers of nuclear war, but devoted comparatively little attention to changing international institutions, such as a disarmament verification agency, mechanisms for the peaceful settlement of international disputes and expanding the role of UN peacekeeping forces.

The environment movement ought to learn from that error. Whatever your cause, it is a lost cause if there is not international institutional reform.[15]

---

14 Contact person: Harry Learner, 301 East 45th Street, Apt 20B, New York, NY, 10017.

15 One non-governmental organization which has been campaigning in this way is the World Federalist Association; the West Australian

War Control Planners Inc

To conclude, here is a case study of one small US non-governmental organization (NGO) which has set an example for the environmental movement. First, it has selected a number of basic issues and campaigned on them - irrespective of how fashionable such issues may be. Second, it is not a mass membership organization but it still has managed to have some impact.

Throughout history weapons always have been rendered obsolete by the forward march of technology and management skills. Infantry swords were made obsolete by horse cavalry; cavalry was rendered obsolete by gunpowder weapons; today's nuclear weapons are obsolete because they would be suicidal if ever used. But humankind has always been in a cycle of "improving" weapons. The task now is to abolish war itself - and to recycle military personnel, equipment and expenditure for the betterment of humankind.

War Control Planners Inc.[16] has been exploring (along with many other pioneers) the frontiers of a new international order in which all nations will find defence against foreign enemies, and in which knowledge and skills in all professions can be directed towards production of food, clothing, housing, energy, clean air and water, health, education and national security for the people of all nations.

Harriet Kurtz was a minister of the United Church of Christ. Howard is a former lieutenant colonel in the Air Force and an engineer. They got into this crusade in a strange way. After World War II, Howard was working for American Airlines, planning for the first New York-to-Moscow air service. The Cold War intervened. Howard was in Moscow on May Day of 1947 and saw the new Soviet weaponry on display there and he could see the preparations for war. They became permanent students of global politics. In the 1950's they held regular seminars on

---

contact address is: 212 Watts Road, Wilson, WA 6107.

16 War Control Planners Inc., P.O. Box 19127, Washington DC, 20036.

the subject, collecting ideas and drafting proposals. In 1965 Howard lost his job with a management consulting firm. He attributes this to pressure from two defence manufacturers who were clients. They then concentrated full-time on War Control Planners Inc.

The basic technique of War Control Planners Inc. (WCPI) is to take the latest marvels of war and space technology and try to imagine how this same hardware might be used to assist in global peacekeeping. In other words, WCPI is not automatically opposed to new technology. Technology is neutral; it all depends how it is to be used. Back in 1961, for example, the Kurtzes sent a proposal to President Kennedy suggesting how the US might use modern electronics to help all nations protect their borders. For three decades, then, War Control Planners Inc has pioneered the conversion of military facilities to civilian use. It has produced a torrent of proposals.

For example, in the late 1960's, WCPI came up with a unique arms control proposal. It could have been started on its way without the participation of the Soviets, or for that matter of any nation other than the US. Reconnaissance satellites, seismic detection, and other scientific devices are now advanced enough to permit adequate surveillance of all nations making aggressive preparations. This work is done without violating anybody's territorial sovereignty, and without access to anybody's classified information. Therefore, said the Kurtzes, let the United States take the lead in building such a surveillance system, inviting the world to observe how it works. Meanwhile, the US need not interrupt its ordinary defence strategy in any way. Let the US further take the initiative in developing a United Nations type of Command-and-Control system, open to all citizens of all nations and employing no classified information, to demonstrate that such a system could direct a multi-nation fire-brigade force (if any such force existed). Further steps would involve war-prevention-systems conferences, dedicated to working out political controls, so that no world authority could tyrannize over independent nations or fall under the domination of any particular nation. The Kurtzes reasoned that the Soviet Union and China would perceive that a Global Safety Authority would give them greater national security than they enjoyed as independent strategists. They then would join up.

More recently has come WCPI's Global Information Co-operative (GIC). This international project would pool data collected by orbiting satellites and make it available to all nations for a variety of purposes:

Earth resources - With satellites already in orbit, it is possible to observe the condition of soil, growing rates, rainfall, crop diseases and yields. Infrared detectors can spot blight in crops before the farmer knows it is there. An information centre can tell when a nation is headed towards crop failure and when another nation will have an abundance. Similar readings are possible on the size and condition of livestock herds.

Forestry - Along with its crop monitoring, the GIC can keep similar track of the world's lumber resources, spot fires and other dangers more efficiently than ground observers.

Wildlife - Satellites can track herds, plot migration patterns and help protect endangered species.

Mineral resources - Space age "prospectors" can plot new mineral deposits, water and other resources via satellite. They also can locate geological faults unknown even to the governments of the nations concerned and, hopefully, help to head off or at least reduce the impact of natural disasters.

Weather - Even if weather control remains some way off, satellites can provide regular weather data to the 80 per cent of the earth which does not receive reports.

Some of the Kurtzes' proposals are moving towards implementation, such as a UN satellite verification system. But much more remains to be done.

One lesson from this case study is the need for vision - always be ready to be imaginative. Where there is no vision the people perish!

Second, keep plugging away. Even if the public and politicians are not presently interested - they could become so. Do not become discouraged.

Third, fashions come and go. What is not fashionable at the moment - could easily become so. Verification, for example, was a "non-issue" in the early 1980's because the world was gearing up for World War III. Then came the Gorbachev/Reagan INF breakthrough - and suddenly government officials were looking for ideas on how to verify an INF agreement! (At this time, the total US expenditure on disarmament - not just on verification but on all disarmament activities, such as international negotiations and publications - was less than the Pentagon spent on military music!) Quite suddenly the US had to develop an international team of verification experts!

Finally, governmental opinion changes quickly. In Australia, for example, the "greenhouse effect" was dismissed by the Hawke Government one year and embraced by it the next. Having changed the minds of politicians, NGO's need to have specific proposals for action. (In the case of verification, the peace movement has been fortunate to have groups like WCPI doing the "homework"). Far more attention needs to be given to international institutional reform to protect the environment. Perhaps a starting point would be to offer to host the Australian CAMDUN meeting in Perth.

COMMUNITY ACTION FOR ENVIRONMENTAL CARE - MEN OF THE TREES

Barry Oldfield [1]

Richard St.Barbe Baker, was a remarkable man, and a member of the Bahá'í Faith. He was a forester but he had a world view. And he was probably one of the first people in modern times who saw the terrible impact that mankind was having on the planet. While studying at the University in Saskatchewan in Canada, he used to work during the vacations as a lumber jack and was terrified by what he saw in the cutting and the wastage of the Canadian forests. He used to travel around on the wheat farms especially in Alberta in Canada, and again he saw massive loss of soil through overclearing of trees. And he used to tell the farmers they had to put shelter belts back into their landscape. Ten years later, in 1922, he was Assistant Conservator of Forests in Kenya. And there his job was to issue timber felling licenses to the concessionaries to take the teak and the ebony out of those magnificent forests, the tropical forests of Kenya. But he was very concerned that the forests were not despoiled and that they did not take more than the annual yield of those forests. And so he used to go on long safaris into these forests and check the growth of his trees but he was not satisfied just with looking at the tropical forests, he wanted to see all of Kenya. And when he went to the Northern Highlands of Kenya, he discovered a very sad state indeed. There on the fringes of the Sahel, the Kikuyu tribes people were practising a slash and burn type of agriculture. They would cut the scrub, grow a few crops for a year or two and then when the soil had been depleted of its nutrients, they would move on and

---

1 President, Men of the Trees, WA.

cut more of the scrub. And gradually, the Sahel was growing year by year. In some places it was advancing 30 km a year into the Northern Highlands of Kenya.

When he saw this, he said to them, "You must plant trees, you must restore the forest in order to save that soil for future generations." This was entirely a novel idea to the Kikuyu people and so they questioned him on this. And he saw that the way to encourage them to do it was in fact to use the African style of getting geared up for a great occasion. That meant a great dance. In 1922, he inaugurated the dance of the trees and with that began the whole movement of the Men of the Trees.

The funny name that we have is actually a translation of "Watu wa Miti". "Watu wa Miti" is Swahili for Men of the Trees. Men of the Trees are working in 108 countries. His Royal Highness, Prince Charles, is our patron. In Australia, we began in 1979 because of course the whole of Australia is suffering exactly that type of environmental degradation that was a problem to the Kikuyu people in Kenya in 1922. We are losing land through the wastage of human kind. We are not blaming farmers. We love the farmers. They are doing magnificent things. But it is the pressure of the whole of our civilization on the land which is causing massive soil degradation.

In Western Australia, we go out during our winter months, planting trees to try to save that land in our Wheatbelt. We go out every weekend to plant tens of thousands of trees. This year, our total will be close to a quarter of a million trees. It is all done by volunteers. Volunteers, like me and like you, giving up a weekend or two weekends, travelling out and putting our hands in the soil to plant trees.

Since we started tree planting in 1979 the experience of planting trees has taught us a great deal about our earth. You see, most of us live very comfortably in a city. Everything is brought to us in great luxury. But we do so at the expense of the world and its natural resources and particularly the agricultural lands. When we went to the country, we put our hands in the soil, and started to discover things about ourselves, a about the way we lived and about how we have to adjust ourselves to living in peace and harmony with that land. When you dig a hole in sand, and that is all it is out there in the Wheatbelt, you may find that although it is pouring with rain, the soil just a quarter of an inch under the surface is absolutely powder dry: at the same time you are feeling the rasping wind on your cheek and it is blowing that sand off the surface of the soil and hitting your face. We have seen tractors bogged in the middle of the paddock because of a little light rain, and they cannot pull their heavy ploughs, we have become aware of the salt coming up, because the crystalline salt makes the ground so slippery to walk on that you slip and fall. We have realised that this is our country and we have made it like this. Then you start to realise that you ought to do a lot more than just pass a few pious motions and write a letter to a member of Parliament.

It comes down to actually rolling up your sleeves. Beginning with that first experience, we have continued to explore every avenue to fire people's imagination to get them out there planting trees, not paying others to do it. Come and do it yourself. Because this is the way you discover how the world works.

One of the first things we dreamt up was the farm tree project. Previously, we were planting trees along roadsides, but we realised we really needed to plant trees in the paddock: we needed to plant trees to help the farmers on their land. Farmers, of course, need trees which are productive, trees which produce fodder for their sheep and cattle, trees which can produce nuts and fruits and alternative crops. Trees which could even enhance our small floriculture industry. Trees which would produce timber and similar products in a few years time. So we were looking not just at restoring native vegetation, although that has always been a key part of our work but we were also looking at introducing new crop trees to diversify our agriculture. Our farm tree project also aims to put many, many more people on the land. You see, we have depopulated our country in the name of economics. We have told farmers, they get big or get out. And all over our wheatbelt, towns are dying because there is not the population to support local tradesmen and local industries.

And so, again we find that there is a great need to take people back into the country. And the way we can do it is to say, come and join us, come and plant trees.

Another scheme that we have dreamt up is what we call the farm tree help scheme. Under this scheme, people living in the city are linked with a particular farmer, one family in the city linked with one family in the country. The family in the city ask the country family what kind of trees they would like. The farmer may say I would like *Eucalyptus camalduensis*, or I would like some acacias. And so the city family grow 600 trees in their back garden and one weekend, in the winter, they pack those trees up in the car, drive to the country and plant them with the farming family on the farm. Not only does this give the farmer a few extra trees, not many, just a few, but it improves city-country understanding. And that is very, very important. We all believe in putting people and trees together - not trees and machines but trees and people.

Another scheme which looks like blossoming in 1991 is our Million-Tree Day. For the last few years at any rate, we were concerned that Arbor Day has been trivialised. It has been a matter of, give the kids a half day off, we will read a few poems about trees and plant a flowering shrub outside the headmaster's window. Our response during this Decade of Landcare has been to link city and country schools in the same way as we link city and country families. We had a pilot project at Lesmurdie High School in Perth which grew 4,000 trees. Students themselves grew the trees, not the teachers or the gardeners. Their names were on the trays with the trees in it. They grew 4,000 trees and on Arbor Day last year, they took them to Cunderdin, 150 kilometres east of Perth and with the children of Cunderdin Agricultural College, planted those trees on land which had been degraded through soil erosion and from salt encroachment. Now it does not take a degree in higher mathematics to say that if we had 250 schools paired in that way, and we can do that very easily, we shall in 1991, plant one million trees in one day. What a marvellous target! And we're working towards that.

CONSERVATION OF THE ENVIRONMENT

Phil Gabriel [1]

I would like to begin this short paper by endorsing Barry Oldfields' comments (see previous paper) in regard to community tree planting exercises; he has emphasized a very practical approach for the community to get involved in environmental issues, in particular, in the management of agricultural and urban areas through reafforestation. This is a principle which Friends of the Earth (Perth) support.

Friends of the Earth are an international organization. We have dealt with various large issues, such as the nuclear issue, toxic wastes and pollution, whaling and deforestation especially in areas such as South America and Africa over the last 25 years. Friends of the Earth in Perth concentrate mainly on urban issues. At this stage, we are a small organization, and we also consider that it is more pertinent for us to concentrate on urban issues, given that there are a large number of organizations, thankfully, that are interested in the whole spectrum of environmental issues from non-urban, such as forest and agricultural issues, to the urban issues.

We are essentially a lobby group: we lobby government, in particular, to review the policy and planning aspect of the environment from resource development to waste management and food production. We also lobby private organizations to produce particular goods and services that may be environmentally sound in their long and short term impacts on the environment. We take the view, recently summarized by David Suzuki, that the world's resources are finite and the human race is going to exhaust those resources beyond their recovery rate, if it

---

1 Formerly Project Officer, Friends of the Earth, Perth, WA.

continues with its current material exploitation. However, given the unique position that humanity holds, in the earth's biosphere, humans have the opportunity to exploit those resources in ways that do not risk large scale degradation of the planet. Humans have gone from being one part of the world's food chain to a modifier of the food chain and as we all know, we now also have the capability of destroying it and therefore, ourselves.

Friends of the Earth believe in ecologically sustainable development, that is human actions which do not threaten the assimilative capacities and biodiversity of the biosphere. We promote activities such as, greater technological efficiency in energy use, making greater use of renewable resources, such as solar and wind energy, and placing a higher price on non-renewable resources such as fossil fuels. Indeed we believe that there needs to be an overriding valuation of resource use inversely proportional to their impacts on the environment; that is, those which when used cause least impact on the environment have the lowest price tag, provided they ar really available.

On the local scale, we take a permacultural approach to urban management and lifestyles. Permaculture is derived from two terms; permanent and agriculture, which together implies sustainability. The term advocates an approach whereby people perhaps grow more of their own food, maximise the re-use of the waste products from their own households and communities (e.g. to produce fuels such as methane to provide electricity, and fertilizers for vegetable production), use solar technology, (it is actually available and needs further promotion and investment from those who hold the reins of power and wealth) and also recycle waste water (e.g. after use for watering of one's own garden) and so on. We are interested in the development of autonomy or a 'closed loop' for energy and material uses in our urban living and work situations.

We therefore encourage consumers into more appropriate and perhaps simpler ways of living. We present them with practical information where possible in such areas as materials recycling, appropriate cleaning agents and also to educate them into not using known agents of environmental degradation such as disposable nappies. For example, in the United States in 1989 there were 18 billion disposable nappies in land filled sites, which is about 4 million tonnes of waste. So we try to educate mothers, for example, to using cloth nappies and laundry services which have a far less deliterious impact on water resources and land resources in general. We would like to see, for example, sanitary land fill reduced and some time in the future not used at all as a method of waste disposal in Australia.

On a larger scale, we lobby state, local and federal governments for action on economic, energy and conservation policies and practices. We believe that the various government levels have an integral and a very important part to play in encouraging sustainable development by instruments such as tax incentives to businesses that are involved in recycling and other organizations involved in practices that are promoting a more sustainable development ethos. We believe that these businesses and organizations will eventually produce goods and services which, in the long term, are economically viable. We are also confident that consumers in this country and beyond will recognize through an ongoing learning process, that it is more sensible to treat the environment as something which supports us not as something which has an infinite capacity to absorb waste and pollution. Then the market place will determine that the more ecologically sound businesses are worth supporting. By contrast, at present the price of such things as recycled paper is still about 20 percent higher than paper produced through the chlorine bleaching process. We believe these prices will come down if the community perceives that the external benefits of using recycled paper have a worth over and above that of the product itself. As suggested by our permacultural ethos, when people realise that our links to the environment are not simply left outside when we close the door at night to have our evening meal, that we are actually part of the biosphere and essentially linked to it, then that relationship will be one of the principle determinants of our economy.

In summary, we believe that the environmental consciousness in this country, western countries and other countries is probably at an all-time high. It is facilitated by a more unified world approach, the formation of many multi-national committees, and the adoption of treaties and agreements in regard to atmospheric pollution, deforestation and desertification, for example. We believe that the world is now in transition phase whereby governments and communities are having to face up to the cost of past environmental mistakes. We believe that with greater support from the nations, particularly western nations that hold and manage the majority of financial and material wealth, we can create and invest in an environmentally sound and environmentally sustainable future.

SCIENCE, TECHNOLOGY AND THE ENVIRONMENT

Philip Jennings[1]

I want to discuss briefly the importance of technology in our society and the need for all of us to be aware that technology has both benefits and adverse effects: adverse impacts on our society in economic, social and environmental aspects.

Before I start, I have to provide a definition, because it is important to distinguish between science and technology. Science is basically a cultural activity, that people are engaged in, trying to understand the behaviour of the natural world, how the world actually works. We observe it and we try to explain it. A technologist is a far more practical person who is concerned with trying to invent new things often to make a profit, sometimes out of a desire to help the human race or some of the other creatures of the earth.

Our society, over the last 200 years, has been increasingly influenced by technology. Probably if you go back to about the middle of the eighteenth century, that was the time that humans began to realise that they could greatly improve the quality of their lives by using technology; by inventing machines, many of which were driven by fossil fuels, such as coal and oil and more recently, gas. These machines were able to do a lot of the back-breaking labour that humans had to do prior to that and gradually all areas of our lives have become influenced by, if not dominated by technology. So it is most important that we understand the nature of this technology because technology, if used wisely, can greatly enrich our lives and make it possible for us to build the sort of sustainable

1 Associate Professor of Physics, School of Mathematics and Physical Sciences, Murdoch University, Murdoch, W.A. 6150.

society that we all want. Technology, if used unwisely, can bring great pain, hardship and even destroy the sort of world that we are living in.

Let me explain what I mean by technological progress. Politicians are very fond of talking about technological progress as though it is an extremely desirable goal. In its broader sense, technological progress just means the development and implementation of new forms of technology. These could be weapons of mass destruction, they could be factories to produce new food products, or they could be new transport systems, all of which have benefits and problems associated with them. I think it is important to understand that technology always comes at a price and what I will do here is list some of the prices.

In a paper as short as this, I cannot go into great detail, so you will have to just accept some of these assertions or you can challenge them later, but these I believe are the fundamental aspects of technology.

The first one is **that technology has both beneficial and harmful social and environmental impacts**. Any form of technology you can think of has those sorts of impact. Obviously we would not be interested in the technology unless it has benefits but there also is a price to be paid, both in social and environmental terms.

In fact, we can quantify that more by saying that **all technology is subject to a law of diminishing returns**. Thus although a little bit of a technology may be good for you and more may be even better, if we go on developing that technology, we will reach a point where the costs start to outweigh its benefits and we eventually reach a stage where the further use of that technology becomes inappropriate. For those of you who are more mathematically inclined, let me just show you that in the form of a graph (Figure 1). Those of you who have studied economics will know all about the law of diminishing returns as economists use it. The law of diminishing returns applies in a far wider field than just economics. It also applies in social and environmental terms to the technology we use.

Let us think of a few examples. We all think that automobiles are wonderful for getting us around. However, if you reach a stage where

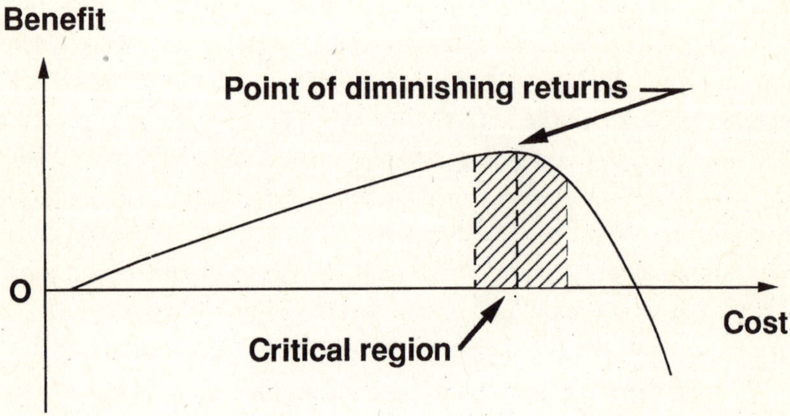

Figure 1 The Law of Diminishing Returns for Technology

there are too many automobiles in a city, you get caught in traffic jams and people start wondering if it is really worth the effort. You can build more roads. If you start building more roads, then the quality of life deteriorates, the noise and air pollution increases, so there is obviously a critical area where you have to evaluate your whole transport philosophy.

Look at agriculture. Fertilisers are wonderful things. They help to improve crop yields. However, if you go beyond the optimal use of fertilisers, you start polluting the waterways, you start causing the eutrophication of lakes and estuaries and it costs millions of dollars to clean them up, if in fact, that is even possible. These are examples, you can think of hundreds more, of the law of diminishing returns. Technology used in moderation, used sensibly, can have great benefits for us. Go beyond a reasonable point, and you get to a situation where the costs begin to outweigh the benefits.

Well, what do you do if you get into these situations? Generally, what our society tends to do is look for a technical fix. Probably the greatest technological fix of our time was President Reagan reaching for the strategic defence initiative. He reasoned that since nuclear weapons were bad, and they were threatening the survival of the human race, the USA should develop a system that was going to make them impotent and obsolete. And so he went for the ultimate technical fix, to build a shield against nuclear weapons which, if you think about it, is so easy to sabotage. Since President Reagan has gone, the strategic defence initiative is disappearing very rapidly as well. That is on a grand scale but it also applies on a minor scale, when for example say you have an ailment, and you buy a drug, a pharmaceutical, to try to cure it. Many times the pharmaceutical will be effective but if you take the pharmaceutical in excess or over a prolonged period of time, even within the recommended dose, it always causes side effects, some of which can be unpleasant. So, **no technological fix for any problem is ever perfect**. And we have to keep in mind, that **we cannot hope to go on indefinitely curing the problems of technology with more technology**.

A couple more characteristics of technology just to complete its picture. **All technological products become obsolete - they are all subject to a life cycle of public acceptance**. Think about the latest stereo you bought five years ago, or the latest home computer you bought a few years ago, or the latest model of car that you may be buying now: you have to realise that they go through a life cycle as shown in Figure 2. When a new model comes onto the market, it becomes popular, reaches a peak of public acceptance and then it dies away. So when we talk about new industries providing lots of jobs and things like that, we have to realise that all forms of technology go through these phases. It might provide new jobs. It might provide new benefits, but a time will come when these are no longer with us. If we start a new mining operation we dig a hole in the ground and take out the minerals. It provides jobs for a few years and then what are we left with in the end? We have to be aware of the consequences of technological progress. Many times, politicians speak of technology as though it is going to be a permanent benefit, and I suppose, when your perspective is three years, anything longer than that seems like forever.

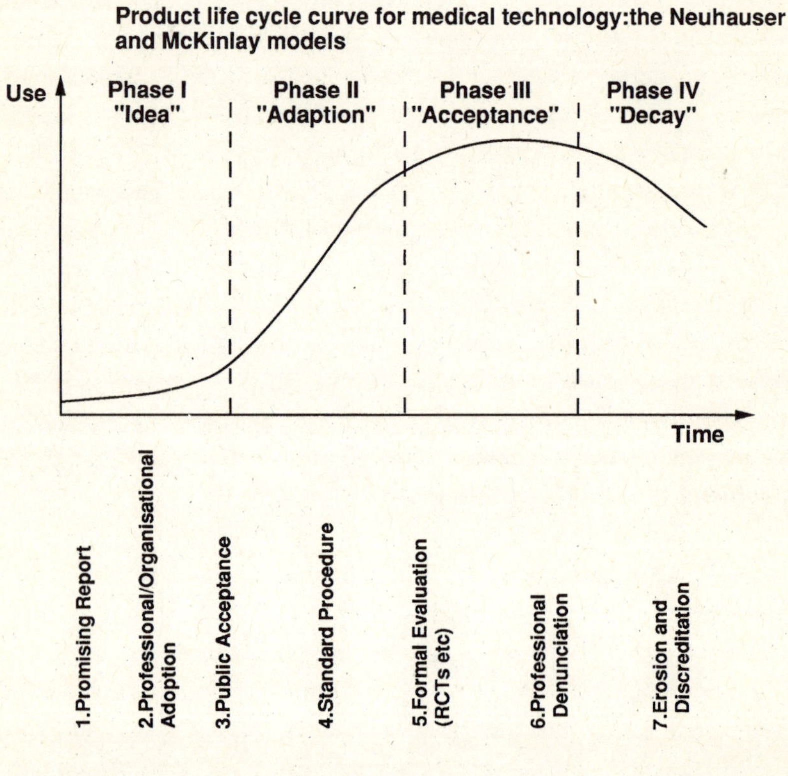

Figure 2 Typical Product Life Cycle

So we come to the final point, the main one that I want to make. It is that, **all new forms of technology need to be very carefully assessed in order to ensure that we maximise the benefits to society and minimise the cost or the adverse impacts**. The adverse impact can be economic, social or environmental. This means that there is a role for all of us in this process. The way in which it is done in our society is

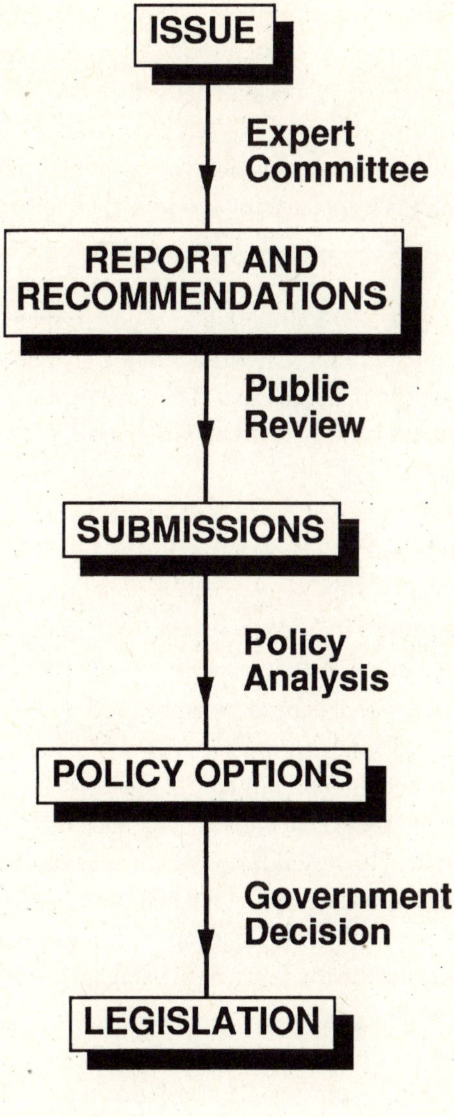

Figure 3 The Process of Technology Assessment

generally a stepwise procedure as shown in Figure 3. In this process, we all have a part to play. Many people do not want to get involved in assessing technology because they feel that they are not technically competent. But that is a mistake because technology has social and environmental impacts as well as the economic and technical aspects and I think that it is important for you to realise that in this process, there is an important role for the public. There is an important role also for people who have scientific expertise. Say there is a proposal for a new plant. The proposal will be put to a government agency which would generally put it out for review and in that process experts and the public have a chance to make an input. This then results in an assessment which goes to government and it leads to a policy and ultimately to a new development.

The reason why we should be involved is that technology is just so pervasive and so important in our society. One of the first people to clearly articulate this concern was E.F. Schumacher who stated that the mark of a mature industrial society is its ability to choose wisely from amongst the range of technology which presents itself.

This concern is further emphasised by Lord Mountbatten who in his famous Strassburg speech in May 1979 stated that science offers us almost unlimited opportunities, but it is up to us, the people, to make the moral and philosophical choices and since the threat to humanity is the work of human beings, it is up to man to save himself from himself. He was talking, of course, of the nuclear threat and I think, the experience of the eighties has been a great cause for hope to us, of how people took their own destiny in their own hands and have drawn back from the brink of that nuclear abyss.

And finally, the words of Albert Einstein in 1945, when he first heard about the nuclear attacks on Hiroshima and Nagasaki:

> We must never relax our efforts to arouse in the people of the world and especially in their governments an awareness of the unprecedented disaster which they are absolutely certain to bring on themselves unless there is

> a fundamental change in their attitudes towards one another as well as their concept of the future. The unleashed power of the atom has changed everything except our way of thinking.

What he was saying is we must all become involved, particularly those with scientific and technological training in ensuring that the use to which technology is put is for the benefit of our society and the benefit of the whole natural world. And therein lies an obligation for everyone of us.

Further Reading

H. Alpert, "Science, Technology and the Future." J. Vacuum Science and Technology, Vol. 18, 1981, 143-147.

K. R. Hammond and L. Adelmann, "Science, values and human judgement". Science, 194, 1986, 389-396.

G. Hefter and P. Jennings, Understanding Science and Technology, Perth: Murdoch University, 1990.

B. Martin, "The naked experts". The Ecologist, 12, 1982, 149-157.

CONSERVING THE ENVIRONMENT - COMMUNITY AID ABROAD

Bruce Wilson [1]

Introduction

Community Aid Abroad (CAA) commenced in a retirement village on the outskirts of Melbourne. Starting as "Food for Peace" it has moved from handouts to self-help, and from development to sustainability and hopefully onto the integrity of creation. Its mode of operation is to forge partnerships between the Australian organisation and organisations in the recipient countries. The indigenous organisations which receive the monies raised in Australia are working with the marginalised, the poorest, and normally the powerless of their society. Historically small community based groups have been the recipients of the funds raised by CAA. The early catch-cry was community-giving-to-community.

In Australia, CAA raises funds in a number of ways; the main ones being:

- the Walk Against Want
- AWARE (a monthly donation scheme)
- CAA community-based groups
- Work Against Want

CAA also supports development education within the Australian community and is a member of the Australian Council For Overseas Aid (ACFOA). Community Aid Abroad publishes a quarterly newsheet, the CAA Review.

---

1 Information Technology Department, West Australian Water Authority, Leederville, W.A. 6007.

The partner organisations in recipient countries present projects for funding. CAA's Projects Department manages the selection, funding and review of projects from the Australian end. A projects committee, comprising volunteer members of CAA recommends projects to the National Executive for approval. These two volunteer-based bodies develop new initiatives in project work. Through the group structure CAA sees itself as a grass-roots organisation. Its members, groups and state committees all contribute to new initiatives.

The types of projects supported by CAA and their mode of operation are illustrated by the following two case studies, one in the Solomon Islands and the other in the Phillipines. Both of these projects have a particular focus on environment issues.

Case Study 1: Rainforests and the People of the Solomon Islands[2]

Tropical rainforest covers three quarters of the Solomon Islands, but at the current extraction rate of 10,000 hectares per annum the estimated life of commercially usable forest is 20 years since less than 1,000 hectares is replanted each year (i.e. % of the extraction).

Against this background, the Solomon Islands economy is experiencing 20% inflation and its foreign debt is rising despite the fact that foreign aid accounts for 30% of budgetary expenditure. Timber exports are important to the national economy since they earn $15 million per annum (1/4 total export earnings) and are second only to tuna fishing. Foreign exchange is needed for fuel, manufactured equipment and goods and to service the foreign debt. The logging companies are mainly Japanese, Taiwanese or Korean which provide timber to the Japanese market for concrete formwork, furniture, packing cases, disposable chopsticks etc.

2 Special Feature, CAA Review, Summer 1990, pp. 3-7.

There are three Australian-based timber companies, and the small quantity of timber entering Australia is used mainly in housing and construction. Lever Pacific Timbers, an Australian-based subsidiary of the giant TNC Unilever was the first and largest of the logging companies in the Solomon Islands. After 20 years of logging, following active resistance to their operations on New Georgia, Levers finished logging in October 1986.

> "After 20 years of logging, Levers left Solomon Islands with tens of thousands of hectares of land reduced to a vine cover that will take at least 500 years to grow back, with no ongoing forest-based industry nor any worthwhile community facilities as was originally promised."

The Malaita provincial government placed a moritorium on new logging operations there in 1987. The Premier in an interview shortly after that imposition stated :

> "On the world market, one cubic metre of timber cost almost US$100. From this the landowners are now receiving only US$3.50. I think it would be far better if the landowners received US$20 per cubic metre leaving $80 for the company to cover labor costs, machinery and profit."

Community Aid Abroad's project partner in the Solomons is the Solomon Islands Development Trust (SIDT), a key indigenous non-government organisation. The sort of dilemma faced by development agencies in the Solomon Islands is illustrated by a quote from the director of SIDT:

> "Unlike in Australia, we in the Solomon Islands cannot afford the luxury of locking all our rainforests up in national parks. There are too many people dependent on these forests for their livelihood - and they have been for centuries."
>
> "I understand the good intentions of people in Australia

> and elsewhere who call for a ban on the importing of our timbers - but I think they are misguided. Such a ban would be regarded in the Solomons as a hostile act - because it does not take account of the effect on people who are already very poor, and need cash income."

Now what can the SIDT do, what does it do? It publishes a magazine LINK.[3] Some quotes from the June/July '89 issue:

> "The village of Olive, on Roviana Lagoon in the WesternProvince of Solomon Islands is one of many which has felt the effects of logging company operations."
>
> A local man who was employed by the company says:
>
> "After we've logged an area, the ground is no good for anything. The company takes any kind of tree, even the ones it shouldn't. They don't care about rivers either - now the water is like Milo. I'm a mechanic, and I have seen engine oil being dumped on the ground. The mud and oil kill reef life and shellfish. The plains in this area are logged out, and now the company wants to move inland and down behind Olive."

Rachel Murray, also of Olive, is very concerned about the effects of logging on her life, the lives of her children and the lives of generations to come. "Before the company came, we used to eat shellfish, oysters, emu seaweed and fish from the reef. We didn't need timber for housebuilding because we had plenty of bush material available. We used to make money selling shellfish too," she says.

> "But now oil and dirt have killed the marine life. Sometimes we are sick from eating contaminated fish. There is no more bush material for housebuilding. In the

---

3 LINK, the magazine of the Solomon Islands Development Trust, June/July 1989.

> places where logging has finished, garden yields are very low, and we have to walk a long way to reach good garden land now. Our population is growing fast, so we need more land, not less."

Down the lagoon at Nusahope, where another section of the same tribe lives, Sae Oka spoke out about how logging has affected his community:

> "We used to have shellfish collecting and hunting parties before our feasts, and we always knew where to go for a good catch. Now that area has nothing, because the pollution has killed the shellfish and the logging has spoiled the wildlife on land."
>
> "The loggers have violated tambu (sacred) sites, as well. In one area where chiefs and heroes of our tribe from head-hunting days were honoured, they have scattered the skulls all over the place. There was an agreement that our tambu sites should be protected, but the loggers have taken no notice of it. When they ruin our tambu sites, our evidence of land ownership is destroyed, too."

"In the future it will be hard if things go on in the same way. We'll have to eat bare ground because that will be all that's left. If we give all our resources to rich people, we'll die in the end," Sae Oka concludes sadly.

But what does SIDT do? First, its aims are "to make villagers and landowners aware of what their options are for the future through its Mobile Teams which tour the scattered villages of the Solomons." So SIDT runs its 34 Mobile Teams (of three to four trained villagers) from nine centres, one in each of the seven provinces, with two in Malaita and Guadacanal. It has provided a vital role in promoting self-sustainable village development in the Solomons since 1983. In these past seven years over 3,000 workshops have been conducted by these teams, and have directly touched the lives of more than one third of the nation.

These village-level workshops are on topics such as: development consciousness; disaster awareness and preparation; practical village living which enhance the villagers' knowledge in such subjects as the wise use of resources (such as timber, land and fish), nutrition, sanitation, family planning, appropriate technology, water tanks and tourism.

The SIDT also endeavours to raise awareness of formal and informal political power structures, and thereby of the forces which are exploiting the villagers, and so empower them to become a critical force in the reshaping of their nation. This political awareness is of great importance where politicians have tended to be anything but "servants of the people". Solomon Islands Development Trust stresses the need for villagers to get together and exercise their power and rights to save the national resources of the Solomon Islands for the future.

Case Study 2: Education Towards Ecological Farming in the Phillipines [4]

The basic objective is to promote ecologically sound living in the tropical rainforest areas with special concern for its tribal inhabitants, the Subanens. This project is a follow-on from an earlier two-year project, Ecological Farming in the Tropical Uplands, which was not fully implemented. This education project takes into account the external forces and other realities which hinder project implementation.

Objectives of the current project are to:

1. Facilitate the education of people, especially upland farmers, towards a sustainable lifestyle within the increasingly fragile ecosystems in the area - i.e. ecologically sound farming;

2. Increasing the knowledge and ability of people (upland farmers, lowland farmers and educators) in how they can help prevent

4 Furthur details available from Community Aid Abroad Projects Development, 156 George Street, Fitzroy, Victoria 3065.

further damage to the ecosystems and what they must do to help restore the balance in those ecosystems;

3. Make these facilities available to:
   a. other groups involved in ecology seminars and wishing to have some practical orientation to the theory they are studying;
   b. small groups who wish to spend time in reflection in a setting dedicated to the restoration and maintenance of the ecological balance; (A reflective atmosphere is encouraged at the centres. This is to encourage each participant to become more attuned to nature and the life that abounds around them.) and ;

4. Providing a special course for mothers and mothers-to-be in nutrition, vegetable gardening, and the value and use vegetables (including cooking)

A further and related objective is the continued development of the Eco-farm in Tonggo and the experimental plots in Katipunan.

Security is a continuing problem. Militarisation since June '88 has caused a major internal refugee problem in the area and a fearful Subanen people. However, maintenance of the program is believed to be one of the factors that contribute to a lessening of tension. The staff is willing to work in areas where previously it would not be possible. As the introduction of new faces to the area could be a cause of further security problems the staff take the necessary measures to ensure the safety of all involved. They need to counter adverse propaganda. Their continued presence and the careful introduction of staff will, they believe, defeat black propaganda.

In courses, half or more of the time is to be spent in practical sessions. The participant is to get sufficient practice so that he/she will be able to implement the learning on their return to their land.

Life style in the centres is very simple (even by local standards). Bunkhouse accommodation is provided (sleeping on the floor, no electricity), plus a kitchen and eating area. Toilet and bathing facilities are simple, healthy and ecologically sound. As much as possible participants are responsible for their own cooking and cleaning. Produce of the training sites is available to supplement participants diets, as well as to introduce them to new vegetables and new recipes. Time spent in the kitchen is considered as an integral part of any training session.

The project aims to establish biogas generators for cooking in the villages. This introduces the need for cash to purchase and maintain pigs which are an essential component of the biogas generation. Reforestation is proposed for four sites at present, and is a long term (ultimate) aim of the project.

Target groups for the project are: Subanen farmers (male and female), other upland farmers, church workers and leaders, and priests. The project coordinator is a priest, and the project is endorsed by his archbishop. Funding sought was $70,000 for a two year period.

Conclusion

Community Aid Abroad believes small groups are the basis through which change can be successfully instigated. CAA funds projects that enable people to retain their integrity and their dignity, that is projects that work at the 'grassroots' through local communities.

ENVIRONMENTAL CARE - THE BAHÁ'Í EXPERIENCE

Richard W. Bell [1]

Introduction

As a community dedicated to "carrying forward an ever advancing civilization" [2] that will last for at least a thousand years, the Bahá'í community clearly has a commitment to development that is sustainable. The strength of that commitment is illustrated by the following statement which links environmental care to the core of the individual's faith:

> "The Bahá'í commitment to the environment is fundamental to our Faith." [3]

More importantly the strength of the Bahá'í commitment to the environment is illustrated by the Bahá'í approach to establishing "a more moderate balance between material development and the requirements of the natural world", [4] as reflected in its initiatives relating to environmental management and care.

Briefly stated, the Bahá'í approach is to take a global perspective of the environment and to view its protection in the broadest possible terms: "barriers that limit fruitful and harmonious cooperation among

---

1 Lecturer in Environmental Science, School pf Biological and Environmental Sciences, Murdoch University, Murdoch, W.A. 6150.

2 Bahá'u'lláh, Gleanings from the Writings of Bahá'u'lláh, Wilmette: Bahá'í Publishing Trust, 1952, CIX, p. 215.

3 Bahá'í International Community, The Bahá'í Statement on Nature, New York: Bahá'í International Community, 1987, p. 7.

4 Bahá'í International Community, A Bahá'í Perspective on Nature and the Environment, Mona Vale: Bahá'í Publications Australia, 1986, p. 16.

men and women, whatever their national, racial, or religious background"[5] are included in the Bahá'í view of the environment.

Despite its global perspective of the environment, the projects initiated by Bahá'í communities to date tend to be small scale, local activities which emphasize spiritual values, unity of effort, self-reliance and sustainability.

Bahá'í Initiatives in Environmental Care

1. *The Early Phase*

Until the mid-1980s the Bahá'í community's direct contribution to environmental care was either through *ad hoc* projects or those of a social and economic development nature. One of the more impressive *ad hoc* projects was a well coordinated campaign during the International Year of Youth (1984) which resulted in the planting of 23,000 trees across the United States of America. Elsewhere during the International Year of Youth more modest tree planting projects were completed.

Most prominent among the social and economic development projects were those concerned with education in which both direct and indirect contributions to environmental care were evident.

In India, since establishing a tree nursery in 1984, the Rabbani High School in Madhya Pradesh state has planted 14,000 trees on campus and in its neighbourhood. Their outreach programme helped establish four village nurseries that have produced over 30,000 seedlings. In 1986, their efforts in environmental education and tree-planting were recognized by the Indian Government's Ministry of Environment and Forestry in the form of an award presented to the school by Prime Minister, Rajiv Gandhi.[6] In Haiti, tree planting has been a major activity associated with

5 Bahá'í International Community, A Bahá'í Perspective on Nature and the Environment, Mona Vale: Bahá'í Publications Australia, 1986, p. 16.

6 Bahá'í International Community, "Greening the Earth: Conservation

the Anis Zanuzi School.[7] Through Bahá'í-sponsored tree planting programmes more than 100,000 trees were distributed over a two year period from 1985 to 1987. The tree planting programme was part of the community development outreach by the school and was undertaken in cooperation with international development agencies.

In Kenya, the Bahá'í community has initiated or participated in reforestation efforts, in environmental education and in the manufacture of fuel-efficient stoves. The development of appropriate technologies has also been the focus of other projects including one in Bolivia, where at the Dorothy Baker Centre of Environmental Studies they have been researching alternative energy sources to wood which is an increasingly scarce resource on the high plateau of Bolivia. The Centre has been exploring technologies including biogas digesters for cooking, lighting and heating, and solar energy use. In addition they have been investigating soil and water conservation measures and distributing training materials in Quechua and Spanish. The Centre is dedicated to providing training opportunities for the indigenous Quechua-speaking peoples.[8]

The development of Bahá'í radio stations in Panama, Ecuador, Peru, Bolivia, USA, Liberia and Chile has also illustrated a Bahá'í commitment to the innovative use of appropriate energy sources and construction materials in order to establish the stations with low cost in rural environments.

In Tasmania, the Kentish Bahá'í community has staged three very successful "Alternatives" weekend events. These events were held to present to the surrounding rural communities alternative lifestyles and agricultural practices.

Whilst the above brief survey of conservation projects has high-

---

work ranges worldwide". One Country, 1(3), 1989, pp. 6-8.

7 Bahá'í International Community, "Private organizations play a key role in Haiti's reforestation". One Country, 1(4), 1989, pp. 10-11.

8 Bahá'í International Community, "*Greening the Earth: Conservation work ranges worldwide.*" One Country, 1(3), 1989, pp. 6-8.

lighted the large size activities, in aggregate terms the smaller more numerous projects probably have had a larger impact on conservation of the environment. However, details of such projects are difficult to obtain. Indeed, the difficulty of obtaining details of the numerous small grassroots projects illustrates an important dilemma - namely that they are largely invisible. By contrast with Government-sponsored projects or those conducted by large corporations, grassroots projects are not supported by large public relations arms which widely publicize their achievements. Consequently the wider community is often oblivious of the substantial achievements of grassroots projects. The number of grassroots organizations working on development projects including those devoted to environmental care exceeds 140,000.[9] Their collective achievements are substantial but often unreported. Innovative methods are urgently needed to bring these achievements into the public arena as a source of hope and inspiration for the future and as an antidote to the gloomy negative images portrayed in the mass media.

2. *The 1980s onwards*

The Bahá'í community's commitment to the environment was given a higher profile in the 1980's through several related initiatives beginning in 1983 when they were urged to engage in Social and Economic Development projects for its own benefit and that of the wider community as a way of articulating the Bahá'í vision in practical terms. Within four years, there were nearly 1,500 such projects underway, mostly concerned with education.

Within the Bahá'í context, Social and Economic projects are a natural outgrowth of needs within the community to achieve "a dynamic coherence between the spiritual and practical requirements of life on earth."[10]

---

9 A. B. Durning, "*Action at the grassroots: Fighting poverty and environmental decline*." Worldwatch Paper, 88, 1989.

By contrast with many attempts at social and economic development, the Bahá'í approach acknowledges that success depends on grass-roots initiative and direction in projects which should have as their aim the upliftment of the community and the promotion of self-sufficiency and self-reliance. Consequently, the projects undertaken have tended to be small scale. Moreover, through promoting the acquisition of skills in consultation as a fundamental tool for action the projects have facilitated widespread community participation in the projects, including that of women and minority groups.

In promoting the engagement of Bahá'í communities in social and economic development, both short and long term outcomes were envisaged. Improvement in the material well-being of communities would be a tangible measure of their benefit. More importantly, the long term value of the learning experience gained from such activities would enhance human honour and self-esteem, community resourcefulness, flexibility and cohesiveness. The above principles also guide Bahá'í communities in their environmental projects.

Two years later (in 1985), as its contribution to the International Year of Peace, the Universal House of Justice (the international governing body of the Bahá'í community), released its "Promise of World Peace"[11] which outlined its vision for world peace and offered the Bahá'í experience as a model for study to reinforce hope in the unity of the human race. The relevance of the Bahá'í experience as a model for environmental action by communities will be examined later.

Thirdly, in 1987, the Bahá'í International Community was formally admitted as the sixth member of the World Wide Fund for Nature's (WWF) Network on Conservation and Religion. In so doing, two state-

---

10 Bahá'í International Community, A Bahá'í Perspective on Nature and the Environment, Mona Vale: Bahá'í Publications Australia, 1986, p. 16.

11 Universal House of Justice, Promise of World Peace, Mona Vale: Bahá'í Publications Australia, 1985, p. 22.

ments were published which for the first time articulated a Bahá'í view of the Nature and the Environment.[12,13]

In its 1987 statement on Nature, the Universal House of Justice stressed that environmental problems all require a transnational approach. Their prognosis for a resolution to environmental problems is predicated on a new political order based on "principles of economic justice, equality between the races, equal rights for women and men, and universal education".[14] Furthermore, their analysis concludes that "the major threats to our world environment....are manifestations of a world-encompassing sickness of the human spirit,....mastered by an overemphasis on material things and a self-centredness that inhibits our ability to work together as a global community" [15]

Fourthly, in its 1989 statement to the Bahá'í world, the Universal House of Justice called upon communities to make conservation of the environment an integral part of their ongoing activities by "...assisting in endeavours to conserve the environment in ways which blend with the rhythm of life of our community." [16]

They gave concrete expression to this call by the establishment in 1989 of an Office of the Environment attached to the Bahá'í International Community's office in New York. The Office of the Environment defined its task as providing a link between environmental expertise at the international level and the growing body of community-based environmental projects world-wide. It will focus on trees - tree planting and

---

12 Bahá'í International Community, The Bahá'í Statement on Nature, New York: Bahá'í International Community, 1987, p. 7.

13 Bahá'í International Community, A Bahá'í Perspective on Nature and the Environment, Mona Vale: Bahá'í Publications Australia, 1986, p. 16.

14 Bahá'í International Community, The Bahá'í Statement on Nature, New York: Bahá'í International Community, 1987, p. 7.

15 Bahá'í International Community, The Bahá'í Statement on Nature, New York: Bahá'í International Community, 1987, p. 7.

16 Universal House of Justice, Message to the Bahá'ís of the World, Ridván, 1990.

protection, regeneration of woody vegetation, and conservation education and training.

At this stage of its development, the direct contribution of Bahá'í communities to environmental care has been relatively modest. However, the foundations have been established from which a comprehensive contribution can be made. These communities, over 30,000 worldwide, have been inculcating a set of values, and acquiring a set of skills which equip them to contribute more actively and effectively in environmental management in the future. The value of the Bahá'í model and its potential to contribute to environmental care will be further explored below.

The Bahá'í Experience - A Model for Study

To conclude this paper, I wish to explore further the notion of the Bahá'í experience as a model for study as suggested in the "Promise of World Peace".[17] What are the characteristics of the model and what is its relevance to those interested in the future of the planet? The present account of the Bahá'í model for change draws substantially on that outlined by Khan,[18] to which the reader is referred for further details.

The Bahá'í community is an evolving organism: its literature is replete with expressions of dynamic change and evolutionary growth; communities expect to evolve; individuals expect to grow and mature in capacity; and their experience of its short history reinforces that expectation.

For the individual, the anticipation of personal growth is coupled with the injunction to modify personal behaviour by individually enquiring after truth. It is an injunction that commits the individual to a lifelong, open-ended learning process. Since the search for truth occurs within a

17 Universal House of Justice, The Promise of World Peace, Mona Vale: Bahá'í Publications Australia, 1985, p. 22.

18 Janet Khan, "The Bahá'í Community sa a Model for Social Change", Presented at 31st Annual Meeting of Int Soc. for General Systems Research, Budapest, Hungary, Bahá'í News, No. 682, Jan. 1988.

supportive and tolerant community, the individual has not only consent but also encouragement to grow and develop. In the Bahá'í community, for personal growth is through active involvement in the life of the community the mechanism in all its diversified forms embracing worship, fellowship, and service. Active involvement in the complete spectrum of community activity provides a wide area for the individual to develop their full human potential. This service produces tangible outcomes in terms of social and economic development and at the same time is a vehicle for spiritual growth of the individual. By contrast with some religious communites which promote asceticism as the path to spiritual enlightenment and growth, in the Bahá'í community the individual is encouraged to develop spiritually through service: the act of service is elevated to the rank of worship.[19]

A positive attitude to change empowers the individual since its corollary is that individuals can change their present and future circumstances. That process of empowerment is vital since positive attitudes provide the will and aspiration which facilitates the discovery and implementation of practical measures.[20]

For the Bahá'í community there is a similar expectation of dynamic change. There is an awareness that with time, the Bahá'í community needs to turn its attention to more issues in the wider community. It envisages developing its own capacity to deal with these issues. Apparent crises such as those posed by environmental degradation can within the Bahá'í context be viewed as a challenge to be met rather than a source of despair. In meeting the challenge the community develops a new maturity and capacity for coping with further challenges. In formulating solutions to environmental damage, it can count on the enthusiastic support of the community's membership. By contrast, many

19 'Abdu'l-Bahá, In: A Compilation on Bahá'í Education, Oakham, U.K.: Bahá'í Publishing Trust, 1976, p. 83.

20 Universal House of Justice, The Promise of World Peace, Mona Vale: Bahá'í Publications Australia, 1985, p. 22.

other communities operate on the expectation that plans for change will elicit resistance and produce polarization among their members. Consequently, the basis for decision-making in the political context is to minimize opposition or electoral damage. Such an approach to decision-making may not be consistent with policies for sustainable development since the latter often requires short term sacrifice and long term planning.

A second element of the Bahá'í experience as a model for study is its global framework. Over 30,000 communities around the globe co-exist in a network that provides a global perspective on priorities for action, open and rapid exchange of information globally and explicit encouragement for local autonomy and action. Communities are encouraged to focus on local issues and problems and to generate appropriate activities to alleviate constraints. Through the interchange of information and personnel on a regional and global scale there is a protection against parochialism in dealing with local environmental issues. In the developed countries such changes need to take account of the need to decrease consumption.[21] By contrast, in the less developed countries, the alleviation of poverty deserves priority. Both issues must receive equal attention and appropriate response from the Bahá'í community.

In the context of environmental care, the Bahá'í model indicates that individual Bahá'ís would be receptive to the suggestion that changes in lifestyle are needed. From the above elements of the Bahá'í model we can predict that to the extent that cooperative community efforts are required to solve environmental problems, individual Bahá'ís would respond by participating actively. The extent to which Australian Bahá'ís might be motivated to act for the common good in environmental care is illustrated by a recent study of three sample groups: expatriate Iranian Bahá'ís, Australian Bahá'ís, and unselected Australians.[22] By contrast

---

21 Bahá'í International Community, The Bahá'í Statement on Nature, New York: Bahá'í International Community, 1987, p. 7.

22 N. T. Feather, R. E. Volkmer and I. R. McKee, "A comparative study of the value priorities of Australians, Australian Bahá'ís and expatriate

with a sample representing the general Australian population, the Bahá'ís of both Iranian and Australian backgrounds demonstrated a much higher levels of altruism. Further studies of this sought could prove extremely valuable as it would provide empirical evidence on the extent to which the efficacy of the Bahá'í model could be shown to be universal. More importantly, it would emphasis the nexus between environmentally sustainable development, the ethical principles that derive from the Bahá'í scriptures, and the model of social order being developed by Bahá'í communites and environmentally sustainable development. The relevance of such studies is highlighted by the emphasis given to the "ethics of sustainability" in the current draft of "Caring for the World", the World Conservation Strategy for the 1990's.[23]

Further Reading

Office of Social and Economic Development, Bahá'í World Centre, Social and Economic Development, Mona Vale: Bahá'í Publications Australia, 1988, p. 48.

Research Department, Universal House of Justice, Conservation of the Earth's Resources, Oakham, UK: Bahá'í Publishing Trust, 1990, p. 26.

Iranian Bahá'ís", J. Cross-Cultural Psychology, (In Press), 1991.

23 D. Munro and R. Prescott-Allen, Caring for the World - A Strategy for Sustainability (2nd draft), Gland, Switzerland: World Conservation Union, United Nations Environment Program, and World Wide Fund for Nature, 1990.

ECONOMIC GROWTH VERSUS ENVIRONMENTAL DAMAGE: A GLOBAL PERSPECTIVE

Fariborz Moshirian[1]

Introduction

Global environmental damage is proceeding at an alarming rate. The major forms of environmental damage were outlined above by Newman (this volume). Deforestation is a particularly significant form of environmental damage with global implications and will be discussed in some detail in the present paper.

> "Tree cover is one of the most visible indicators of the earth's health and, because trees are an integral part of basic life-support systems, they are most vital. On sloping land, the loss of trees can accelerate rainfall runoff and increase soil erosion, diminishing land productivity and aggravating local flooding. Where tree cutting exceeds regrowth, deforestation releases carbon that contributes to the build-up of atmospheric carbon dioxide and a warming of the earth."[2]

According to Park[3] environmental problems can be examined at four levels:

Local: many policies must deal with environmental problems at the sub-national level. The three most common units of scale are

1 Lecturer in International Economics, School of Banking & Finance, The University of New South Wales, Kensington, NSW.

2 L.R. Brown, W.V. Chandler, A. Durning, C. Flavin, L. Heise, J. Jacobson, S. Postel, C. Pollock, L. Starke, and E.C. Wolfe, State of the World, New York: W.W. Norton & Co., 1988.

3 C.C. Park, Environment Policies, London: Great Helm, 1986.

topographic (e.g. a large drainage basin), ecological (e.g. a large forest), and political (e.g. a state, county or regional district). Such policies should logically reflect the broader national policies, modified where appropriate to take account of local factors and constraints.

National: policies are required to cope with environmental problems which arise in individual countries as a result of that country's own activities. Examples include land use, development control and natural resource management. Such policies are also required to deal with pollutants which are emitted into the environment of one country and which are not diffused abroad.

Transfrontier: some environmental problems transcend national boundaries, and it is appropriate to deal with them bi-laterally or multi-laterally. Transfrontier policies are required where two or more adjacent countries are linked by an environmental system, and activities within one country affects environmental quality in the foreign countries. Examples include the release of sulfur dioxide from the United Kingdom and Germany which ultimately falls as acid rain over Scandinavia; and the use of the Rhine River by France, Switzerland, Germany and the Netherlands.

International: some environmental problems (especially air and water pollution) are truly international in character. Sometimes these relate to a particular area (such as the Mediterranean which receives pollutants from rivers in all surrounding countries). Often, however, they are global (such as rising levels of carbon dioxide in the atmosphere). Many of the most urgent problems are international because political segmentation and difficulties in national objectives and ambitions prevent consistent inter-regional implementation of global environmental management strategies. While I acknowledge the importance of the activities of those who are working at the local and national levels to reduce environmental damage the purpose of this paper is to analyse:

- the underlying factors contributing to environmental damage at the transfrontier and international level;

- world security in the context of the environment as opposed to military security;
- some of the international agreements which could have saved the environment of this planet in the 1970's and 1980's; and
- some of the environmental benefits that can be derived from a united world.

Factors Contributing to Global Environmental Damage

Fossil Fuel and Environmental Damage

According to Brown *et al.*[4], the consumption of fossil fuel is releasing into the atmosphere over 5.5 billion tons of carbon dioxide every year. This volume of carbon emitted into the atmosphere is the main contributor to the Greenhouse Effect and hence the challenges of this generation are : (a) to raise the efficiency of current energy generation, distribution and consumption and (b) to develop feasible alternative sources of energy. Hydro-power, power from the sun, power of the wind and nuclear energy have been identified as the potential future energy sources. However, the amount of money invested in this area, relative to its potential contribution to sustained economic growth and a safe environment is very small. Between 1950-1979, while the world economy grew on average by 4 percent per annum, the consumption of fossil fuel use worldwide quadrupled and yet it was only after the oil crisis of 1973 that the Organisation for Economic Cooperation and Development (OECD) countries initiated a serious programme on research and development (R & D) to raise the efficiency of the current

4 L.R. Brown, W.V. Chandler, A Durning, C. Flavin, L. Heise, J. Jacobson, S. Postel, C. Pollock, L. Starke, and E.C. Wolfe, State of the World, New York: W.W. Norton & Co., 1988.

energy use and find the feasible alternative sources of energy. According to the International Energy Agency,[5] the governments of OECD countries invested $5 billion in R & D in 1975 for their energy programmes. As prices of oil continued to rise, they increased their R & D to $10 billion by 1980. However, as the price of oil started to fall during the eighties, their R & D expenditure on energy fell back to $7 billion by 1986.

Deforestation and Trade Imbalance

Deforestation, the second main factor contributing to environmental damage,[6] is releasing about 2.5 billion tons of carbon dioxide into the atmosphere. The underlying factors contributing to deforestation are (a) trade imbalance between developed and developing countries and (b) population growth. Research by Houghton *et al.*[7] indicates that the major areas experiencing deforestation are tropical America, tropical Asia and tropical Africa. Not surprisingly, countries in these regions are facing severe economic problems which are caused partly by lack of opportunity to export their commodities to developed countries. In an attempt to earn foreign currencies to finance their industrial development projects, and ease their balance of payment difficulties, the countries of these regions are heavily engaged in the export of the timber, one of a few natural resources that they can still export to the world market.

---

5 International Energy Agency, Energy Conservation in IEA Countries, Paris: OECD, 1987.

6 L.R. Brown, W.V. Chandler, A Durning, C. Flavin, L.Heise, J. Jacobson, S. Postel, C. Pollock, L.Starke, and E.C. Wolfe, State of the World, New York: W.W. Norton & Co., 1988.

7 R.A. Houghton, R.D. Boone, J.R. Fruci, J.E. Hobbie, J.M. Melillo, C.A. Palm, B.J. Peterson, G.R. Shaver, G.M. Woodwell, B. Moore, D.L. Skole, and N. Meyers, "The flux of carbon from terrestrial ecosystems to the atmosphere in the 1980 due to changes in land use: Geographic distribution of the global flux," Tellus, 39B, February/April 1987, pp. 122-139.

Population Growth and Environmental Damage

The third factor contributing to environmental damage is population growth. The rate of world population growth is about 2 percent. If this trend continues, by the end of the century the world population will be 6 billion as opposed to 3 billion in 1960. The growing population in developing countries, the shortage of foreign currencies and inadequate technology prevent these countries from raising the productivity of their agriculture sectors as fast as growth in the rate of their population. One of the consequences of this failure is the creation of additional arable land for cultivation at the expense of forests in these countries. While this strategy is providing a short term food solution for the growing population of these countries, in the long run, they are faced possibly with accelerated soil erosion, reduction in the productive capacity of their land and ultimately its possible desertification.

Consumption of Fuelwood and Environmental Damage

The oil crisis of the 1970's negatively affected the economy of most developing countries. High oil prices pushed developing countries to borrow substantial amounts of capital partly to finance their imports of oil which in turn led to a Foreign Debt Crisis in the late seventies and early eighties. On the one hand, the oil crisis led the majority of people in developing countries to face a shortage of the energy required for the expansion of their agricultural sectors and, on the other hand, it increased their dependence on fuelwood as the main source of energy. Indeed, more than two-thirds of all developing countries rely on wood for cooking and heating. Rural dwellers depend on wood almost exclusively. In many countries- including most of Africa - wood not only dominates household energy use, it provides more than 70 percent of energy used for all purposes. Actual planting for fuelwood has averaged about 550,000 hectares annually, a fifth of what is needed.

Environmental Security versus Military Security

Given the above factors contributing to environmental damage, one can see that future world security is in danger if humanity does not respond to environmental damage with a united effort and adequate financial and scientific resources. While the true security of this planet is under threat, the nations of the world are concentrating on maintaining their political security at the expense of the security of the planet itself. The current worldwide expenditure on armaments is $950 billion dollars per year as opposed to $7 billion dollars per year by the governments of OECD countries on R & D for alternative energy sources. While environmental protection requires a large number of scientists to raise the efficiency of the current energy use and increase the chance of using alternative energy sources, over 500,000 scientists are working in the military sectors against each other to maintain world political security. Military R & D is heavily weighted towards development, meaning that most funds go into building prototypes and making demonstrations, not into basic sciences which are far more likely to lead to new inventions and commercial products. According to Brown *et al.*[8] less than 3 percent of U.S. military research has gone into basic science. The military share of U.S. Federal R & D funds increased from roughly 50 per cent in 1978 to 68 percent in 1988. Furthermore, while Japan and Germany allocate 2.5 per cent of their GNP to non-military research expenditures, the U.S. share is only 1.9 percent. The $3.2 billion allocated to the U.S. Strategic Defence Initiative in 1987 was 20 times the federal energy conservation research budget, which declined by almost 7 per cent. In developing countries where population growth, balance of payment difficulties and low agricultural productivity are contributing to global environmental damage, military expenditure is much higher than the amount of money spent on health and education (Table 1). In such a political environment, where countries are spending more on their military sector as opposed

---

8 L.R. Brown, W.V. Chandler, A.Durning, C. Flavin, L.Heise, J. Jacobson, S. Postel, C. Pollock, L. Starke, and E.C. Wolfe, State of the World, New York: W.W. Norton & Co., 1988.

to other sectors of the economy, one wonders to what extend world security in the context of environment can really improve ?

TABLE I. Weighted Average Shares of Central Government Expenditures by Type of Economy, 1982 (%).

Type of economy	Defence	Education	Health	Social Welfare	Economic Services	Other
Low Income	**18.5**	**5.5**	**3.0**	**5.0**	**25.2**	**42.8**
Lower middle income	**14.2**	**13.7**	**3.7**	**6.8**	**23.5**	**38.1**
Upper middle income	**11.5**	**10.9**	**5.1**	**21.0**	**20.8**	**30.7**
Inustrial market	**13.9**	**4.8**	**11.7**	**40.4**	**9.7**	**19.5**

Source: World Bank, World Development Report 1985, Table 26.

International Crisis versus International Agreements

Given the above contrast between military security and environmental security, this section of the paper draws the attention of readers to three international agreements which directly relate to international problems facing humanity. Unfortunately, despite the economic and environmental difficulties facing humanity, the current international institutions have generally failed to implement the international agreements which could have arrested environmental damage at the global level.

General Agreement on Trade and Tariffs

The GATT (General Agreement on Trade and Tariffs) is technically an agreement rather than an organisation. There are three aspects to GATT. The first is the agreement itself, establishing standards for the commercial policies of the contracting parties. Two standards are fundamental (with a few exceptions): quotas are prohibited, and each nation must observe the Most Favoured Nation (MFN) clause with all contracting parties. The second aspect of GATT is its role in the settlement of trade disputes between nations. GATT has no enforcement machinery, but it does provide an impartial recourse that has been of considerable use in the past. The third aspect of GATT is its sponsorship of tariff reductions. These reductions have been accomplished in a series of multinational negotiations, or rounds. The first two GATT rounds, held during 1949-51 lost the opportunity for removing trade barriers in agricultural commodities due to the concern of a few powerful countries, in particular the United States, which saw agriculture to be a national security issue and hence argued that it should not be negotiated at the global level. That failure and the subsequent formation of Common Agriculture Policy of European countries, generated disincentive among the farmers of developing countries for their export of agriculture commodities. While there are other factors contributing to the poor agricultural productivity of developing countries (such as the emphasis given to industrialisation at the expense of agriculture, inappropriate national development strategies, land distribution problems etc.), the message for the current environmental damage is that if humanity could again seize, with open eyes and global vision, the opportunity for trade negotiation in agricultural commodities, developing countries' agricultural productivity could be increased implying less pressure on forests of these countries.

New International Economic Order

The New International Economic Order (NIEO) arose when groups of developing countries came together to demand, and sometimes attempt, a restructuring of the world economy. The frustrations and aspirations of these countries were voiced in resolutions of the United Nations General Assembly, and given further expression in 1964 at the first U.N. Conference on Trade and Development (UNCTAD), attended by 122 nations - about 90 of them developing countries. UNCTAD was subsequently made into a permanent agency of the United Nations. From such sessions and meetings there finally emerged in 1974 a demand for the NIEO. The more important proposals in this program include the following:

1. An integrated program to stabilise and support the prices of a group of Primary Products in terms of manufactures,

2. increase development aid (aid to slow down population growth, increase education of people, improve agricultural productivity etc.),

3. the expansion in the developing countries of various manufacturing activities now located in the developed countries and

4. the transfer of advanced technology from the developed countries to the developing countries.

Now, 16 years later economists and officials believe that NIEO is a dead issue. In other words, the world never responded to the implementation of the NIEO call. Obviously one of the consequences of this failure was further pressure on environmental damage from two directions: (a) poor economic development of Least Developed Countries (LDC) contributed substantially to continuous population growth and (b) lack of expansion of export sectors of LDC's meant further reliance on the export of timber. The lesson which can be drawn from the NIEO is this, many factors which could have prevented current environmental damage were raised in the 1970's, not in the context of environmental damage, but rather in terms of economic development. Hence failure of the world to accelerate economic development in the seventies implied further environmental damage in the eighties and now in the current decade.

Desertification[9]

In September 1977, the United Nations Conference on Desertification (UNCOD) agreed on an impressive plan of action, largely directed at governments, to tackle desertification. The plans called for a wide range of activities to deal with the complex biological, social, economic and political factors involved, based on the development of proper land use, including conservation and enhancement of living resources and water resources. It envisaged the successful halting of desertification throughout the world by the end of this century through national programmes assisted by inter-governmental and non-governmental organisations coordinated by the United Nations Environment Programme (UNEP). This formidable cost, estimated at $15,625 million per year, was seen as coming from increased levels of international aid and reinforced if possible by arrangements such as a special anti-desertification fund and an international taxation scheme.

Since then progress has been disappointingly slow. According to Mohammed Kassas, former President of the International Union for Conservation of Nature and Natural Resources (IUCN), neither the governments of the countries affected by desertification, nor the rich donor countries who were expected to provide substantial funds, have yet shown any serious commitment to the action that is desperately needed. Attempts to increase the flow of multilateral aid have stalled. This may be partly due to lack of organisation by countries suffering desertification. Each was supposed to set up an official focal point to coordinate its anti-desertification activities. In many of the countries, however, the communication gap between government departments is so great that one might be preparing programmes to prevent desertification while another is setting up agricultural and pastoral projects that will increase it. Coordination is clearly vital, but not a single African gov-

9 Discussion on desertification in this paper is based on R.Allen, How To Save the World, Bernes and Nobles Beeks, 1980.

ernment has set up a focal point. Partly, however, donor governments appear reluctant to give more money through UN organisations, even though recipient governments often prefer the absence of strings attached to multinational aid.

Having discussed the above three international agreements, the next section attempts to find a solution for the current environmental damage.

Environment in a United World

Given the above international agreements which could not provide a concrete change in the structure of the world economy and hence the global environment, one wonders what went wrong and/or why humanity could not respond effectively to international crises. The answer can be found in the report of the World Commission on Environment and Development entitled "Our Common Future" which in its first chapter emphases the fact that "the earth is one but the world is not one"[10] While this prescription is the key issue which should be considered for the security of the planet, I can remind you that such a solution was prescribed over 120 years ago by the prophet-founder of the Bahá'í Faith, Bahá'u'lláh, in His address to kings and rulers of the world when He proposed a system of collective security. In His message He placed emphasis on the significance of unity by saying that "the well-being of mankind, its peace and security are unattainable unless and until its unity is firmly established"[11]. In the other part of his message he says "so powerful is the light of unity that it can illuminate the whole earth".[12] In other words, before the publication of "Our Common Future" and

10 World Commission on Evironment and Development, Our Common Future, Oxford: Oxford University Press, 1987.

11 Bahá'u'lláh, Gleanings from the Writings of Bahá'u'lláh, 2nd ed., Translated by Shoghi Effendi, Wilmette: Bahá'í Publishing Trust, 1976, p. 286.

12 ibid, p. 288.

other recent books which stress the significance of a united world, the Bahá'í Faith not only placed emphasis on the unity of mankind as a prerequisite for the security of this planet, but also it has already established a United Community in over 140 countries which can be used as a model for the establishment of a united world. Given the severe environmental damage facing humanity, there are few choices left and it is up to this generation to choose the most feasible of these. These choices are as follows:

1. To accept a united world concept and work towards establishment of the unity of mankind, as the Bahá'í Community has been doing for over 140 years.

2. To accept the current political systems, where arms expenditure (over $950 billion dollars per year) and the employment of over 500,000 scientists in military related projects are diverting the necessary resources required for the restoration of environmental security, and continue to generate financial and scientific resources for the restoration of environmental security through cut-backs on health, education and social welfare budgets in western countries. In other words, to accept a lower living standard for the preservation of this planet.

3. As a means to save this planet, continue encouraging the consumers in western countries to become more environmentally conscious by relying less on the motor car, using less electricity, and less resources such as metals, paper etc.

Among the above three choices, the second choice is not socially attractive or even feasible in the current political climate. While the third choice may improve to some extent environmental damage in western countries, this choice is not going to stop deforestation in developing countries nor necessarily provide an alternative source of energy or even raise the efficiency of current energy usage. Only the first choice can ensure that humanity as a whole becomes equipped to restore in a logical, feasible and socially desirable way, the security of the planet. Once the first option is adopted, then the third choice can be considered as a complementary policy to accelerate the restoration of the environmental security of this planet.

Let us see the possible environmental benefits, at the international level, which can be derived in a united world.

Technological Changes in a United World

One of the most fundamental changes in the structure of the world economy, once the unity of mankind is established, will be removal of the military sector (an international policeforce which should remain to ensure the internal and external order of a united world) implying that the world economy will have an additional $950 billion dollars to spend every year on other sectors of the economy plus an additional 500,000 scientists who will join their colleagues in civilian sectors. In a united world, the world scientists would no longer work against each other, but rather work cooperatively to bring about the technological changes required for sustainable economic growth. The resource reallocations resulting from the transfer from a divided world, maintained by military forces, to a united world, maintained by global cooperation and scientific collaboration will have, among other benefits, the following impacts on environmental security of the planet:

1. There will be more financial assistance and more scientists for raising the productivity of the agricultural sectors in developing countries. This progress will ensure that less forest will be destroyed.

2. Population growth will slow down implying that there will be a significant decline in new global environmental damage. The population crisis is not a technological challenge but rather a resource reallocation challenge which can only take place in a united world. The human financial resources and existing technology can be mobilised in a united world at the international level to slow down growing population.

3. Instead of R & D on military weapons, the scientists of the world will be asked and encouraged to conduct R & D on alternative sources of energy, implying that for the first time in history, they will be working on a united front to overcome the energy crises. Furthermore, there will be more financial resources available in a united world to continue raising the efficiency of current energy use. The benefits of such resource reallocation will be a substantial change in the economic life of billions of peoples, a significant slow down in consumption of fossil fuel and hence less environmental damage through the greenhouse effect. There are many more benefits which will be derived from a united world such as an improvement in international financial markets, better macroeconomic policy coordination to reduce the risk and uncertainty of the current world economic system etc. However, these issues cannot be discussed in the present paper.

Natural Resources in a United World

The second main benefit of a united world is protection of natural resources including forests, in developing countries. Under the current social and political system, each country has absolute sovereignty over its raw materials and natural resources. In such a political environment, the environmental movements may be able to stop logging of the forests in Australia. However, when concern about the forests in Brazil and many other developing countries is raised, there is not much that people in western countries can do. Note that while in western countries people are becoming very conscious of their environment and they are working hard to protect it, the main deforestation is occuring in developing countries, and hence if in western countries the environment is to some extent protected, the rest of the world, as result of economic hardship, will continue destroying nature. Thus only a united world can bring about the following two changes in this respect : firstly, the economy of a united world will minimise economic pressures on developing countries through the transfer of some financial and technological resources from the developed countries and will provide better trade relationships between developed and developing countries to decrease the dependence of the developing countries on the export of timber. Secondly, in a united world where International Institutions have authority to monitor the resources of the world, the volume of trade in timber will be managed on a sustainable basis. The implication of this international policy is that the importers of timbers (mainly America, Japan and European countries which can no longer destroy their own forests) will be forced to conduct R & D, to find substitutes for timber and substitutes for timber products such as paper. The point is that, when the importers of timber are faced with the abundant supply of timber in the international market (in an economic sense, faced with a perfectly elastic supply of timber, as has been the case over the last 100 years), their scientists and their politicians are not going to direct a substantial R & D effort into finding substitutes for paper and other similar goods. To demonstrate the capacity of scientists to respond to an international crisis when the right policies are placed in operation, consider the oil crises of the 1970's. During that time, some economists held the view that the high price of oil creates balance of payment difficulties, increases inflation and ultimately reduces the living standards of people in western countries. However, as the high price of oil reduced the volume of imported oil to those countries the scientists between 1975 and 1985 were able to raise the efficiency of energy use quite substantially. In the United States for instance, the energy efficiency of the motor car has increased by 92% during that time. If the scientists are able to raise the efficiency of an existing technology by 92% in ten years, humanity should be more confident that scientists can do much better for the current energy crisis and consumption of fossil fuel when the world is united and a global goal and vision is given to its scientists. In other words, there are many missing opportunities in the scientific fields for this generation, simply because the world is divided in its political, scientific and economic policies.

Global Environmental Protection in a United World

There are a number of economic activities which negatively affect the global environment such as international water pollution, acid rain, emission of chlorofluorocarbons etc. which require international cooperation. However, under existing laws, it is very difficult to control international pollution. If acid rain is damaging the forests of the Scandinavian countries, French and German producers cannot be penalised. If in Eastern Europe, economic growth is the main national priority without any environmental consideration, pollution of rivers in that region will continue without any compensation to other affected regions of Europe. In this divided world where most laws and regulations are confined to national activities, global environmental damage cannot be effectively reversed. Similarly, in the absence of concrete environmental regulations in developing countries, the economic activities of multinational corporations will continue damaging the global environment. The above environmental damage can be controlled in a united world through an International Tax and Subsidy Policy which will control the polluters and compensate the polluted.

To elaborate on this point let us consider the impact of regulations imposed on private companies and factories during the 1970's and 1980's in Western countries. During the 1970's when governments of Western countries became conscious of the pollution problem, they imposed some restrictions on production levels of factories in the form of tax, to abate the pollution level. The argument against these regulations was that since the producers were forced to produce less in order to maintain a permissible level of pollution the consequence of these regulations would be slower economic growth, a rise in unemployment and consequently, lower living standards. However, a recent OECD analysis[13] indicates that the regulations led neither to unemployment nor

13 Organisation for Economic Cooperation and Development, Environmental Policy Benefits: Monetary Valuation, Paris: OECD,

to economic difficulties but rather it created incentive for producers to come up with innovations to maintain their level of production and economic activity by reducing and controlling pollution. Now the reason why humanity is incapable of reducing global environmental damage is because the current divided world cannot impose and reinforce regulations such as an international tax scheme which would be able to reduce environmental damage and in turn bring about those innovations and technological changes required for a sustained economic growth. In other words, humanity is missing, in the absence of a united world and International Institutions, many opportunities for technological changes and innovations which are so desperately required to save this planet.

Conclusion

While protection of the environment at the local and national levels can be partly dealt with in isolation from the rest of the world, the restoration of environmental security at the global level requires a united world. In other words, the prerequisite for the environmental security of this planet is consciousness of the oneness of mankind. There are a number of environmental benefits which can be derived from a united world. Agricultural productivity of developing countries will rise, thereby slowing down deforestation; the consumption of fossil fuel will decrease as more scientists, released from military research, devote their capacities and resources to overcome the energy crisis. Under the supervision of international institutions, natural resources including forests will be protected implying an acceleration in the research for substitutes for wood and paper. On the one hand, trade relations between developed and developing countries will improve, leading again to less deforestation and, on the other hand, environmental damage to international waters, rivers, Scandinavian forest, etc. will be minimised, through

---

1989.

the international tax schemes which reward countries for decreasing their generation of pollutants.

THE ENVIRONMENT AND THE GROWTH OF SPIRITUALITY

Ray Meyer[1]

Some scientists have tended to ignore the notion of religion as being 'revealed' knowledge. Rather, they put their trust in the theories of science and acknowledge that there are aspects of human experience that are difficult to quantify. Most would also agree that it is difficult to study 'spirituality'. One problem is that if an observable product can be quantified, the very process of quantification can destroy or limit the spiritual experience. Yet spiritual leaders like the Buddha, Jesus or Bahá'u'lláh and the experiences of most, if not all people from the beginning of recorded time, testify to the spiritual nature of man. Educators and teachers usually avoid experiences to develop spirituality for many reasons, one being they tap into belief systems that evoke an emotional response, especially from parents. There are proofs of the existence of God as a source of spirituality but these are unconvincing without intuitive knowledge of God. It is a waste of time, for instance, proving to an infant that his mother loves him. He knows. It is the same with a believer who has accepted or experienced God in his life.

Nevertheless, it will be useful to endeavour to penetrate at least a short distance into the spiritual nature of reality.

During this century it has been fascinating to observe the search for the smallest particle of matter. Right up to the end of last century it was thought by the scientifically aware that the atom was the smallest particle in the universe even though it had not been actually observed using the most powerful microscopes. In 1911 Ernest Rutherford deduced from available evidence that the atom was not the smallest

1 143 Tuross Boulevarde, Tuross Heads, NSW 2537.

particle but was made up of a positively charged nucleus around which a number of negatively charged electrons orbited. The nucleus was subsequently shown to have protons which were positively charged and neutrons which have no electrical charge. Now protons have been further broken down to quarks, quarks into flavours and colours and still the fundamental particle of the universe eludes us. With the movement of all particles, however, energy may be emitted as waves smaller than the wave length of light.[2]

Whether the energy that moves the smallest particles or the greatest stars and galaxies is spiritual, is still one of the mysteries that science has not solved. Both scientists and theologians, from their different viewpoints have struggled with this problem. Many scientists, and Einstein is reportedly one, have retained their belief in God at the same time as they have pushed their knowledge and scientific theories to the limits. Why not? What Einstein proved mathematically was known intuitively by a Persian mystic centuries ago.

Split the atom's heart and lo!
Within it thou wilt find a sun.[3]

Christian theologians and philosophers like Kierkegaard, Tillick, Barth and Robinson, [4,5] have reconciled science and religion by seeing God as the Spiritual Essence, the God Within, the Ground of Being, the Depth of Life. The concept of a creating God, up there, or out there, caused Christian believers to hold on to myths and dogmas that were opposed to scientific enquiry and, in many cases, opposed to the teachings of Jesus. These Christian thinkers have helped twentieth-century

---

2 For a more detailed discussion see: S.W. Hawkins, A Brief History of Time, Sydney: Transworld Publishers, 1989, p.66.

3 Bahá'u'lláh, The Seven Valleys and the Four Valleys, Wilmette: Bahá'í Publishing Trust, 1986, p. 12.

4 A development of these ideas may be followed up in: F.C. Happold, Religious Faith and Twentieth Century Man, Ringwood: Penguin Books, 1966, Chapters 6, 10, 11.

5 J.A.T. Robinson, Honest to God, Adelaide: SCM Press, 1963.

believers to realize that they are not bound by an Old Testament cosmology that simply is not true in a literal sense. The clash between Christianity in all its main forms and science on issues like evolution and existence in an expanding universe, had to occur.

Scientists like Teilhard de Chardin who are also Christian theologians, reconcile science and religion by enlarging still further our understanding of man, the universe and the creative forces that unite all.

> When, in the universe in movement to which we have just awakened, we look at the temporal and spatial series diverging and amplifying themselves around and behind us like the laminae of a cone, we are perhaps engaging in pure science. But when we turn towards the summit, the totality of the future, we cannot help engaging in religion.
>
> Religion and science are the two phases of one and the same complete act of knowledge - the only one which can embrace the past and the future of evolution so as to contemplate, measure and fulfil them.[6]

Teilhard sees man as being part of the spirit of the earth. Spiritual energy is found in the trillions of cells making up each individual and forms a unity with the whole universe. While the mind of man is now becoming conscious of the immensity of the universe, the forces of growth and of evolution are impelling him towards the planetization of mind and his unity with all. He also sees that love, the affinity of being with being, is a general property of all life. It is pure energy. For Jesus, love is the creating power of God and is summed up by the affirmation of John "God is Love".[7] Almost 2,000 years later in the revelation of Bahá'u'lláh we find:

---

6 Pierre Teilhard de Chardin, The Phenomenon of Man, London: Coliins, Fontana Religious Books, Seventh Edition, 1969, p. 312.

7 The New Testament, 1 John, 4, Versc 16.

> O SON OF MAN!
> I loved thy creation, hence I created thee. Wherefore, do thou love Me, that I may name thy name and fill thy soul with the spirit of life.[8]

And in the Bahá'í Writings:

> Know thou of a certainty that Love is the secret of God's holy Dispensation ... the fountain of spiritual outpourings. Love is the vital bond inherent ... in the realities of things ... the living link that uniteth God with man, that assureth the progress of every illumined soul. Love is the most great law ... that bindeth together the diverse elements of the material world, the supreme magnetic force that directeth the movements of the spheres ... Love is the spirit of life ... The establisher of true civilization.[9]

Environment

In our approach to understanding the environment we will assume firstly the spiritual unity of all existence and secondly, that there is enough consistency in relationships between particles of matter or even between human beings to enable behaviour to be predicted.[10] We will accept that there is a life-force in the universe and that all living things share in it.

The seed of the spotted gum tree will grow into a spotted gum tree through the life-force within it. This force is more than the encoding of the genes although height, shape, colour and beauty are potentially there

8 Bahá'u'lláh, The Hidden Words (Arabic), London: Bahá'í Publishing Trust, 1975, p.6.

9 'Abdu'l-Bahá, Selections from the Writings of 'Abdu'l-Bahá, Haifa: The Universal House of Justice, 1978, p.27.

10 See also on these matters, S.W. Hawking, The Uncertainty Principle, op. cit., Chapter 4.

in the seed. The environment enriches or detracts from its growth. With a human being, the sperm and the ovum, once fertilized, grow into a human being. The life-force within it ensures its growth, the environment enhances or limits it. The life-force within all living things is real, silent and unseen. But if anything tries to stop its flow, resistance is encountered. Its drive is always towards fulfilment - the actualization of what it is meant to be.

Before we look more closely at how the growth of spirituality in human beings is affected by the environment, we will define the environment and examine its scope.

Is the environment that which surrounds; that which is outside? It is meaningless without a point of reference - outside of what? The individual? The atmosphere we breathe, the vegetation that surrounds us, the animal life with which we share the planet, the minerals, the hills and plains and valleys, the land and ocean, the sun, moon and stars, the solar system, the galaxy, the universe? The things that man has made? The houses, cars, machines, instruments, works of art, open-cut mines, deserts, denuded mountains, polluted seas? Certainly the physical composition of the planet forms part of the environment and also the universe of matter of which the world and its people form a part.

When it comes to human life, however, the social environment could be more significant to man than the physical environment. The mother in whose womb we grow. Father, brothers, sisters, grandparents, aunties, uncles, teachers, friends, community, the nation, the world of people, those who have gone before, those who will follow - all enhance or limit growth to a greater or less degree and add greatly to the complexity of existence. Significant interactions take place between the individual and other people.

What of the environment of ideas? Not all of the environment is visible. Ideas have been accumulating since the beginnings of human life on the planet. They are passed on by word of mouth, translated into actions that can be observed like paintings and dance, translated into implements and used, or works of art that are preserved, or recorded in books, on tapes, stored in libraries. Computers enable people to organize and store information and ideas and recall them in all kinds of new combinations. No human brain can hope to encompass the knowledge we now have. Ideas form a powerful environment, and stimulate the flow of other ideas.

Like the environment of thoughts and ideas, the environment of electromagnetics is not all directly observable, yet the universe is pervaded by it. A probe a million miles in space can be controlled by energy transmitted by waves. Crystals can emit and receive electromagnetic waves. Sound, light, heat, ultra violet rays, infra red, X-rays, lasers, radio and television rays can be generated, controlled, transmitted and received.

Other unseen aspects of the environment are feelings, emotions, values and attitudes. Such components affect the particular person, others specifically, mankind generally and all aspects of the natural environment. They tend to be life-destroying or life-enhancing: hate and love, sorrow and joy, ugliness and beauty, prejudice and openness, fear and trust, selfishness and kindness, destruction and creation. In such aspects of the environment where is the point of reference? It cannot be the physical person. It must be within, perhaps the mind, or the true self or the individual soul. Once we search for the real person other aspects of the environment open up. The trillions of cells that compose the body and the finely balanced systems within the human frame are essential parts of the total environment. Endless interactions are possible between the life-force of the developing self and the environment. The within becomes part of the without. The invisible environment seems to be more than co-extensive with the physical environment. It does not seem to be limited by physical reality. It uses the physical to achieve new ends. By tapping into a particular wavelength a person can materialize in full colour, movement, and sound in a room on the other side of the world or on the moon or on the planet Venus and beyond. Emotions can be stirred. Action can be initiated, reactions induced.

Teilhard de Chardin sees the collective outpourings of emotions, mind, soul, spirit of human beings throughout the planet as forming the **noosphere** which in itself becomes part of the environment, with which individuals and mankind generally, interact.[11]

The Growth of Spirituality

Growth is a partly observable guide to the fact that life-enhancing interactions are taking place.

Dewey defines education as growth which means an increasing capacity to experience and interact with the environment at higher levels of development.[12] Because the stimuli in the environment as we now conceptualize it, present a universe of possibilities for interaction, CHOICE is involved. Each individual becomes characterized by the choices he makes. His own growth and the environment with which he interacts are determined by these choices. By the time a child is in kindergarten his choices are already tending to be life-enhancing or life-destroying. Basic love, acceptance and trust are experienced in the womb. So are fear and rejection. Confidence with other people, delight in beauty have their roots in the pre-school years. The choices the child is already making in these early years help to create the future and determine the kind of adult who will enhance or destroy his own life and the environment he inhabits.

The growth of spirituality involves first of all the acceptance of life. Parents before and after the birth of their child can accept the gift of a new life and experience the joy of creation, for the new born child potentially possesses the attributes of God. Most life-enhancing experiences do not have to be taught at all - they are shared. A little child is one with the beauty of the sunset, one with the waves and the green grass.

Aboriginal people accept their oneness with the animals, the trees,

---

11 Pierre Teilhard de Chardin, op. cit., The Collective Issue, p. 261 ff.
12 J. Dewey, Experience & Education, New York: Collier Books, 1966, Chapter 3.

the rocks, the earth. Traditionally they receive their health and their energy from the environment. Interaction with natural things is spiritual both in the giving and receiving. What they have known in their inmost being is known to all people who recognize their oneness in spirit, who accept that they are part of existence, affected by and in turn affecting the spiritual elements surrounding them.

To an Aboriginal Australian, a "sacred site is a place where Aboriginal people find a manifestation of divine power, a sense of contact with a creative force. In the traditional view, all nature is sacred, but in certain places the spiritual power manifests itself more readily. These are the places where the great events of creation history took place, and the landscape itself is a network of potent locations, linked by the paths of the ancestral creators." [13]

Elder of the Yuin Tribe, Guboo Ted Thomas, is a surviving source of knowledge of the 'dreaming' associated with the sacred mountains on the south coast, Goolaga (Mt Dromedary), the woman's mountain, and Mumbulla, where the young men have been initiated into the tribe, some as recently as 1990. He, and his wife Ann, independently, have been conducting 'dreamings' for Aboriginal and non-Aboriginal Australians for the past ten years, using the energy of the sacred sites, their rich cultural heritage and the beauty of the natural environment of the bush to create understandings between white and black. Guboo has celebrated his cultural awareness with indigenous people in other parts of the world. He claims it is time for a Renewing of the Dreaming. The Renewing of the Dreaming is the re-establishment of the innate spiritual relationship with the earth. He believes that we can all move toward a union with the environment, with each other and with our inner selves. In this way people of all races, can move into the future and continue the spirit of the 40,000 years of human occupation of this continent. We can be the

13 Burnum Burnum, Aboriginal Australia, Sydney: Angus and Robertson, 1988, p. 20.

Nation of the Dawn and lead the world into a new age of unity if we link up with the Aboriginal heritage and release the mighty, harmonizing energies still in this land. Other Australians share this same dreaming.[14]

This consciousness of oneness with nature was expressed powerfully by an Indian Chief, Seattle in 1854. The same ideas have been expressed by Aborigines in Australia who share the same feeling for the land.

> How can you buy or sell the sky, the warmth of the land? The idea is strange to us. If we do not own the freshness of the air and the sparkle of the water, how can you buy them? Every part of this earth is sacred to my people. Every shining pine needle, every sandy shore, every mist in the dark woods, every clearing and humming insect is holy in the memory and experience of my people. The sap which courses through the trees carries the memories of the red man.
> The white man's dead forget the country of their birth when they go to walk among the stars. Our dead never forget this beautiful earth, for it is the mother of the red man. We are part of this earth and it is part of us ...
> The ashes of our fathers are sacred. Their graves are holy ground, and so are these hills, these trees, this portion of earth is consecrated to us ...This we know, the earth does not belong to man; man belongs to the earth.
> This we know. All things are connected like the blood which unites one family. All things are connected.
> When the last red man has vanished from the earth, and his memory is only the shadow of a cloud moving

14 See an excellent account of recent initiations and renewing of the dreaming in: N. Parbury, Survival, A History of Aboriginal Life in NSW, Sydney: Ministry of Aboriginal Affairs, 1986, p. 26.

> across the prairie, these shores and forests will still hold the spirits of my people. For they love this earth as the newborn loves its mother's heartbeat.[15]

The oneness of spirit was captured again by the English mystic, Thomas Traherne, and by mystics of all races and times.

> Your enjoyment of the world is never right till every morning you awake in Heaven ... and look upon the skies, the earth and the air as celestial joys ...
> You never enjoy the world aright till the sea itself floweth in your veins, till you are clothed with the heavens and crowned with the stars; and perceive yourself to be the sole heir of the whole world ...
> Till your spirit filleth the whole world and the stars are your jewels ... The world is a mirror of Infinite Beauty, yet no man sees it. It is a Temple of Majesty, yet no man regards it. It is a region of Light and Peace, did not men disquiet it. It is the paradise of God.[16]

If, as adults, the inherent oneness of existence experienced by children, has not been crushed or destroyed by prejudice, their love, empathy and trust flow out to others and to the world they inhabit. Such people do not destroy life, they seek to enrich it further. They are open to the knowledge now available about the destruction of life and the environment. They will not destroy the beauty with which they identify. They value animal life and respect the forests. They are prepared to sacrifice to enhance the ecology. They seek the further integration of the isolated patterns that are emerging so that the wonder and beauty of the whole of creation becomes clearer and more complete.

The application of such values in the search for the true self is to

---

15 Cited in David Suzuki, Inventing the Future, Sydney: Allen & Unwin, 1990, p. 230.

16 Aldous Huxley, The Perennial Philosophy, London: Fontana Books, 1958, p. 78.

find God "standing within thee, mighty, powerful and self-subsisting."[17] This is the life-force within. The application of such values to the environment is to find God as the life and energy of all things visible and invisible.

The further growth of spirituality Bahá'u'lláh describes in **The Seven Valleys**. It is worth noticing that the most spiritual experiences are still set amidst natural phenomena within and outside the individual. The first is the Valley of Search, and the next of Love, then Knowledge. In the Valley of Unity the seeker "drinks from the cup of the Absolute and gazeth on the Manifestations of Oneness ... He looketh on all things with the eye of oneness and seeth the brilliant rays of the divine sun shining from the dawning-point of Essence alike on all created things and the lights of singleness reflected over all creation." The Valley of Wonderment he enters like a child and is "tossed in the oceans of grandeur ... at every moment he beholdeth a wonderous world, a new creation, and goeth from astonishment to astonishment, and is lost in awe at the works of the Lord of Oneness."

The ultimate is the Valley of True Poverty and Absolute Nothingness where the true lover reaches the "presence of the Beloved and the fire in the lover's heart will kindle a blaze and burn away all veils and wrappings ... and gaze on the Beloved, and forget all else." This station is dying from self and living in the divine. Baha'u'llah then says that the journey need not be long. For the Spiritually aware the Seven Valleys may be seven steps. With others, seven breaths, or even a single breath if God desires it.[18]

The process of creating generates its own energy. The affirmation of oneness is the relationship between the actualizing self and the

---

17 Bahá'u'lláh, The Hidden Words, London: Bahá'í Publishing Trust, 1975, p. 8.

18 Bahá'u'lláh, The Seven Valleys and the Four Valleys, Wilmette: Bahá'í Publishing Trust, 1986.

environment is motivation in itself. Each life-enriching choice creates energy.

Conclusion

The spiritual experiences shared by all men, the knowledge of their interdependence with each other and the total environment provide a fundamental approach to the ecology. It is fundamental not only in that the experiences are in accord with the true nature of man but because it recognizes that the growth of all the elements interact with the growth of the whole. Energy flows from the dynamic process itself as well as from God, the source of all energy. Each Manifestation of God increases the surge of energy. Bahá'u'lláh releases new life for the body of mankind and shows the path to follow to restore and replenish mankind and the natural forces of the planet. In this process much is still to be learnt from children and from such Aboriginal people as are closest to the earth and to the living spirit of God. Real progress with the ecology will not be found in a bit of this and a bit of that but in a world view that sees men as part of the whole creation, alive in God's world. To destroy the ecology is to prejudice the future of mankind.

True lovers of God's creation feel the shockwaves as a mighty tree crashes to the ground. They are sensitive to the waves emitted from the rocks. They are part of the living pulse of the earth. They grow with the rhythm of the universe. They feel the slight shimmer of the leaf stirred by an evening breeze. They suffer with every unjustly treated child and every wasted life. What they know intuitively can be tested, classified and clothed in words and theories, but the mystic knower holds the key to the future of the planet. It is only the values that flow from God - from Love - the Spiritual Reality that preserve, enrich and manifest the true purpose of existence.

The spiritual ground-swell now taking place enables us all to express our true selves as we share the exhilarating task of creating a new future.

THE INNER LIFE AND THE ENVIRONMENT

Ron Price[1]

It would seem only obvious that the inner life of human beings is that critical zone in which all environmental questions ultimately begin. It is the inner and private aspects of our human existence that is the foundation of all outward expression and is acted upon by the environment in often mysterious as well as more obvious ways.

It is my hope in the course of this paper to suggest a vocabulary, engender a perspective, to open up a language, and to reflect with you on just what this inner life is which "moulds the environment and is also deeply affected by it".[2] In the simplest of terms the inner life has to do with what one thinks and feels, the knowing and the loving dimensions, the cognitive and the aesthetic domains. Obviously, too, this inner life is equally concerned not only with the 'what' but the why, the how, the when and where. It is a busy little place.

It is in this inner life that one locates passion and indifference, exaltation of spirit or alienation, the precious and at times terrible moments of despair, the utter helplessness and defeat, the shame and sorrow as well as the joy and happiness of people. These inner feelings and thoughts are at the very centre of a human being's life. Our concern here is with that aspect of these inner experiences which are not shared. They are in a profound sense one's real possessions.

This inner life is, then, a composite. It contains the inner chatter which we all engage in just to get through the day. It contains our attempts

---

1 Lecturer in General Studies, South East Metropolitan College of Technical and Further Education, Thornlie Campus.

2 Shoghi Effendhi,In: Social and Economic Development, Mona Vale: Bahá'í Publications Australia, 1988.

to still the chatter in our moments of contemplation. For some, this contemplation is silent; others contemplate to Beethoven or Jimmy Hendrix. It is here where invention, the power of discovery, and creativity is born. It is here that civilization is created and flourishes; it is here that Michelangelo's deepest preoccupation, to free his soul from matter,[3] took place. It is here that romance begins and marriages go on the rocks. It is here that the environment is polluted or made whole.

Knowing things, wanting things and doing things have their origins in this inner life. How does the human community arouse the appropriate volition, how does it inspire the acquisition of the desired knowledge and how does it get people to do what responsible people should do? To put these questions another way: can the human community make decisions about what should go on in this inner life?

Human societies have been making decisions about what should go on in this inner life for millennia. Anthropologists call the way this process 'enculturation'; sociologists call it 'socialization'; others might call it 'indoctrination' and still others 'education'. Thomas Hobbes, writing in an early phase of this modern age, argued that whoever defines the names and makes the definitions of the human community controls the society.

It is not the intention here to review the history of human society and to focus on the key determinants of the inner life of people. I would like, though, to draw our attention to one of the transformations that have occurred in the last hundred years, perhaps as far back as what is commonly known as the industrial revolution, in the latter part of the eighteenth century.

First the print media and then the electronic media, increasingly became the filter for virtually everything most people knew about anything. The determiners of what went on in people's inner life became a host of strange bedfellows: advertisers, business people, governments

3 Kenneth Clark, Civilisation, Harmondsworth: Penguin, 1969, p. 99.

and oppositions, a burgeoning entertainment industry, indeed anyone or any organization with a message, with news, with something to say, to sell, to sing about.

One and all they vie for a slice of the cake. The cake in this case is the thoughts and consciences of human beings: the inner life. This inner world has become a garbage pail of candy-floss entertainment, the latest Time magazine, bits of pop-psychology, an endless stream of news, weather, trivia, parental guidance, curricula: a veritable cornucopia.

Some critics have referred to this process of the invasion of the private realm as 'the consciousness industry'. Some see the process as essentially liberating, others as imprisoning.[4] The inner life has become a realm of entrepreneurial investigation. It is also an area that gets filled up with information in such a profusion that sustained thought often become impossible and the life of the imagination stultified.

"We are drifting", Charles Fair expressed the process, "into a chaos of private notions amounting almost to superstition."[5] Whether as specialists in professional life or religious enthusiasts in private life we and our society have slipped into an unsettled emotional state. We can scarcely talk to each other. It is all part of a radical subjectivity and a chaos of inner experience. Practical matters take precedence. Education becomes pragmatism: for jobs and the great consumer society. The inner life, in the process, withers, atrophies and becomes more a collection of garbage than a mystic chalice. Nonsense fills the void of the inner life: quick fixes, shortcuts, handy-hints, how-to books, life made simple, the day's news in six video-clips and the weather, nicely packaged.

The great religions, of which the Bahá'í Faith is but the latest,

---

4 Stuart Ewan, Captains of Consciousness: Advertising and the Social roots of the Consumer Culture, 1976; Michael Schudson, Advertising, the Uneasy Persuasion: Its Dubious Impact on American Socity, New York: Basic Books, 1984.

5 Charles Fair, The New Nonsense: The End of the Rational Consensus, New York: Simon and Schuster, 1974, p. 88.

contain certain essential psychological truths about this inner life. Perhaps its most basic truth is that the human being is inwardly divided and can unify itself only at a cost of continual pain and effort. Man alone is self-made. To aspire is not his option but his fate. It is not the freedom to do as he wants that is crucial but the freedom to be what he should be-a human being with inner completeness. This has been a core issue since Socrates and the prophets of the Hebraic tradition in the first millennium BC.

The process has always been difficult. Each stage in man's cultural evolution: tribe, clan, city state, nomadic, hunter-gatherer, pastoralist has posed its own particular challenge to the individual in search of truth, integrity and inner wholeness. Our age has its own especial features which make this process nearly impossible for millions to achieve with so much nonsense around.

When people have abandoned faith for reason they find that the task of living by reason is too difficult. Enlightened selfishness is followed by selfishness outright. Whom the gods would destroy they first make simpler and simpler. Wholeness, or integration of the self, is not a goal but a battle-at best a balancing act, a perpetually unstable reconciliation of forces which, unreconciled, simply tear us to pieces. A silent destruction takes place in individual lives and sometimes a whole society goes down the tube.

Edward Gibbon described 'the triumph of barbarism and religion' in documenting the decline of the Roman empire in the West. This decline, however one argues the case, had an external and an internal manifestation. The same process, this writer would argue, is taking place in our time. The internal manifestation in the inner life is, of course, much more difficult-if not impossible-to document.

This inner conflict is not a clinical disorder. It is the first law of inner psychic life. The second law of inner psychic life is: 'what goes in must come out'. The final law is what we will call the justice principle. By this is meant 'a process of seeing things with our own eyes and hearing them with our own ears.' Relying on experts, on the reliable information sources, on our parents, our friends can often be dangerous. Indeed, we need to fashion vibrant circles of understanding from our collection of fossilized information about so many aspects of human existence. Not until this fashioning and refashioning process reaches a certain temperature, until 'the sea boils', will we find the pure gold and the true pearls of knowledge.[6]

I would now like to expand briefly on these three laws of the inner life. They are not knew ones but they have tended to get obscured in the dust of modern life.

Inner Conflict

"The true lover yearneth for tribulation", Bahá'u'lláh informs us[7]. By this yardstick it would appear that there are precious few of these people around. The lives of the great Ones who founded the historic religions on Earth bear eloquent tribute to this truth. This is not the place to go into the philosophy and psychology of suffering. Indeed we have all been exposed to it in various forms over the years.

There is one aspect of this first law that needs to be emphasized. It is diametrically opposed to the first law of the hedonistic materialism that is at the root of our way of life. The basis of any ostensible stability in our society has its roots in affluence. Surrounded by endless material goods, by triviality and an intellectual environment fit for ten year olds, the real instruments of terror grow cancerously in our inner life, unbeknown to our own selves.

The quagmire of consumerism and the environmental crises find their origins in the recesses of our own inner lives. Radio, television, pulp-magazine and a list of gadgets *ad infinitum*, the bread and circuses

---

6 'Abdu'l-Bahá, Secret of Divine Civilization, Wilmette: Bahá'í Publishing Trust, 1970, p. 109.
7 Bahá'u'lláh, Hidden Words, Kuala Lumpur: Bahá'í Publishing Trust, 1985, p. 20.

of the Romans, have become the soporific of our souls. The problems of ecology begin in our own minds and hearts, one might call the problem 'egology'.

'Egology' is a natural by-product of the consumer society and its hedonistic and cancerous materialism. The struggle for the higher life, for an inner life of meaning rather than distraction, is a heroic one. We get weighed down by our ordinariness, by the mundane, the routine and the endless distractions. Our ordinary self battles with our heroic self.[8] The battle never ends. Spiritual growth, psychic growth, any growth requires tension. That is the first law.

The search for 'ataraxia', for inner peace, which has been one of the essential domains of search throughout the history of western civilization, must take place within the context of struggle. One of the lies of life, criticized by philosophers, poets and thinkers down through the ages, is the view that "there should be no pain at the heart of life." The first law of the inner life highlights to the true lovers of tribulation as those who point us in the right direction.

What Goes In Must Come Out

Bahá'u'lláh says that there is a divine sign in every phenomenon. The sign of the intellect is contemplation and the sign of contemplation is silence. Rene Descartes used to stay in bed in the morning and think; John Maynard Keynes did much of his writing before he got out of bed in the morning; Einstein said his favourite occupation was thinking. Now I'm not advocating you stay in bed every morning, as Keynes did; nor that you spend all your time thinking. We can't all found systems of philosophy, economics or physics.

It does seem, though, that we could all benefit from a lot more thought, serious thought. We might not want to think about mathematics

---

8 Geoffrey Nash, "The Heroic Self and the Ordinary Self", Bahá'í Studies, Vol. 10, p. 28.

or physics. What each of us thinks about, of course, is a very personal matter. But this choice determines the stuff of our inner life from the time we first wake until our heads hit the pillow at night. Our lamps of inner search, or earnest striving, of passionate devotion, of a certain rapture must get turned on. We need to get turned on and tuned in so that we can display ampler manifestations of human achievement.[9] To what do we turn on and what is the number of the station we should tune into?

The place of the mind, of the intellect, of understanding, can hardly be overemphasized. It is not so much that the mind needs to be stilled, but more that it needs to be activated. Hard, conscious thought, often seen as intellectualizing, is at the heart of any genuinely rich inner life. Precision of thought and language can only come about through effort. We must take command over our craft. There are no short cuts.

This is the second law of the inner life: that the inner life must come to be characterized by 'learning and the cultural attainments of the mind'.[10] Indeed a human being's very happiness is crucially linked with this feature of the inner life. "The happiness and greatness" 'Abdu'l-Bahá emphasizes in elaborating on this primary feature of the inner life, "the rank and station, the pleasure and peace, of an individual (consist)...in his excellent character, his high resolve, the breadth of his learning and his ability to solve difficult problems."[11]

Using One's Own Eyes And Ears

In one of His first Hidden Words, Bahá'u'lláh places justice in the context of individual interpretation of truth :

> The best beloved of all things in My sight is Justice...by
> its aid thou shalt see with thine own eyes...and shalt

---

9 Universal House of Justice, Letter to the Bahá'ís of the United States, December 29th, 1988, p. 4.
10 'Abdu'l-Bahá, op. cit., p. 35.
11 ibid., pp. 23-24.

know of thine own knowledge and not through the
knowledge of thy neighbour...[12]

The essence of the entire Kitáb-i-Íqán, summarized as it is in the first paragraph, is also conveyed in a similar context. Man shall never acquire any certitude or peace "unless and until he ceases to regard the words and deeds of mortal men"[13] as a sign of truth. Shoghi Effendi, the Guardian of the Bahá'í Faith puts this third law in the following way: "Man must eschew affectation and imitation, for every man of understanding will instantly detect their loathsome odour."[14] It is not easy to be original, to have an independence of thought. Einstein said he had only one idea that was all his in his entire life. Anything that separates an individual from the mass ultimately tends to be somewhat painful.

The inner life is subjected to so many pressures, complex problems, images and ideas from the environment that it tends to wander endlessly in search of more and more stimulation. It tends to lack focus in a hell of frenetic passivity or activity. A person needs to work out what his job is and stick to it, as Upton Sinclair put it.[15] Or, as Rollo May says in a different context, we need an internal structure of meaning. The structure of our world is defined in our inner life. This is the basis for the structure we give to our own future, the interpretation we give to our past and the meaning we find in the present.[16]

Shakespeare conveys something of the orientation of this inner life in his description of going to bed worn-out from the day's labours:

But then begins the journey in my head

---

12 Bahá'u'lláh, Hidden Words, Kuala Lumpur: Bahá'í Publishing Trust, 1985, p. 10.

13 Bahá'u'lláh, Kitáb-i-Íqán, Wilmette: Bahá'í Publishing Trust, 1950, p. 3.

14 Universal House of Justice, Letter to an Individual Believer, February 10th, 1980, cited in Bahiyyih Nakhjavani, op. cit., p. 18.

15 Ronald Steel, Walter Lipman and the American Century, Boston: Little Brown and Co., 1980, p. 74.

16 Rollo May, Love and Will, New York: Fontana, 1972, pp. 223-245.

To work my mind when body's work's expired;
For then my thoughts-from far where I abide-
Intend a zealous pilgrimage to thee
And keep my drooping eyelids open wide...[17]

We need to work out, to 'intend' just what we want to journey toward in our heads, what work we want to do with our minds. Swimming as we all do in a sea of information, we each need to work out our focus. What is it that each of us can do? What will we do with our inner life? Samuel Beckett asks this question metaphorically in his 'Waiting for Godot'. Indeed, in some ways the entire corpus of that genre of modern theatre known as the theatre of the absurd is a literary documentary on modern man's failure to define, to design, to adequately equip his inner life. The external environment, the environmental crises we face, is a reflection of this inner crisis, this absence of an inner architecture.

Potentiality and anxiety are two sides of this inner crisis. In actualizing our potentialities, in defining our inner life, with any intensity and direction-tension and anxiety are givens. The journey, too, is a slow one. Defining one's inner life, giving an architecture, a structure to the assorted collection of inside chatter, of willing and wishing-the very basis for action-is at the root of our identity as human beings. It is a very private process and to a significant extent we are on our own.

I would like to conclude this paper with a discussion of the poetry of Roger White as an example of the application of those three laws to the artist, the inner life and the environment. It seems to me that the poetry of this provisional poet-laureate of the Bahá'í Faith suggests a context for this inner conflict, this inner anxiety; an orientation to great minds of this and other ages as antidote to garbage and; an independence of thought-the three laws just discussed.

White has produced four books of poetry since 1979. A dominant theme, especially in his "One Bird, One Cage, One Flight".[18] is solitude

17 William Shakespeare, Sonnets, Ringwood: Penguin, 1985, No. 27.

and community. Ostensibly a homage to Emily Dickinson, the nineteenth century American poet, this small book of poems suggests, indeed cries out, that everything is made from privacy of thought and solitariness. It is a testimony to the confrontation with self and to the cries of all romantic artists since the industrial revolution:

> "I don't want comfort; I want God; I want poetry; I want real danger; I want goodness; I want sin."[19]

The path is one of solitude; for it is in solitude that the richness of the inner life is to be found. The reader leans forward to learn from White who places his own frail vulnerability on the firing line. His openness creates a bridge of trust. In his language, his struggles and his joys, White allows the reader to see the texture of our age: a texture of civilization's virtues, an intellectual energy, a freedom of mind, a sense of beauty and especially a craving for immortality.

At the centre of Roger White's philosophy is an assumption that dissidence is a moral and intellectual contradiction to those who would be peacemakers and unifiers of the children of men. The charm of this perspective is present in poem after poem in a delightful and humorous etiquette of expression. It is an etiquette that possesses a tone, a motive, a manner, a mode that takes the faintest hints of life and converts them into a basis for dialogue.

The dialogue White apotheosizes implies a competence in conversation not yet achieved and a relationship between our inner and outer selves not yet attained.[20] White seems to be saying in a host of different ways that the future belongs to that man or woman who is conscious of the solidarity of the human race. Within this framework of the oneness

---

18 Roger White, One Bird, One Cage, One Flight, Happy Camp: Naturegraph Publishers, 1983.
19 Geoffrey Nash, The Phoenix and the Ashes, Oxford: George Ronald, 1984, p. 62.
20 Universal House of Justice, op. cit., December 1988, p. 4.

of humankind, exists a tension between the individual and the community that is the very source of the artist's ethical creativity.

The struggle for the higher life within all of us is made from the same ingredients of which the artist and the hero are constituted. For the higher life is the inner life where all the battles and the victories are ultimately fought. The artist in dealing with his environment, and especially the human beings within that environment, must respond to the capacities of his audience or else his utterance will have no power. His language must be moderate, tempered and infinitely courteous. It is better, from a Baha'i perspective, to remain silent than to respond with dissent, discord and disdain.[21]

We must be on our guard in this discussion of the inner life that the process is not confused with a withdrawal into privacy. We are talking of an inner life that ultimately finds expression in the environment and in the community. A sense of community feeling lies at the basis of any effective ideology. People who have no community feeling are often unwilling to exercise restraint or self-control in the pursuit of their own self-esteem. They choose to pursue superiority in anti-social ways.[22] The community needs norms on which to pass judgement on human beings. But this requires a unified public morality. A powerful community feeling must exist to overcome the alienation and the acids of individualism that are eating at the heart of our society.

At present the notion of a unified public opinion is absurd. A public with endemic illiteracy, feeble-mindedness, child-like mentalities, a people whose lives are a morass of entanglements, whose vitality is exhausted is not a public in which the notion of public opinion can be taken seriously. At best it would seem that public opinion can only grant

---

21 B. Nakhjavani, "The Artist, Seerer and Seeker", <u>Bahá'í Studies</u>, 10 (1981), p. 14.

22 Paul E. Slipansky, In: <u>Freud's Shadow: Adler in Context</u>, Hillbank: The Analytic Press, p. 269.

legitimacy to those who make the decisions and the rules under which society operates.[23]

The collective mind, at present, is at best impulsive, irritable, incapable of reasoning, exercising justice or engaging a critical spirit and an exaggerator of sense and sensibility. Mass education, now about a century old, will in the decades ahead become more effective. Until it does, "until the nerves and arteries of the nation stir into life, every measure that is attempted will prove vain; for the people are as the human body, and determination and the will to struggle are as the soul, and a soulless body does not move."[24]

White's poetry charts the asphyxia of soul, the offence against life at the heart of our materialist culture. His is a massive, but gentle, denunciation of middle-class Anglo-Saxon society and its bourgeois complacencies. To White, hedonism is a denial of the reality of life and, therefore, a crime against humanity. His call is to arise and struggle and overcome our ordinariness. It is a call to heroism.

The community feeling that White calls us to create is not one based on a simpering puritanism, an over-arching self-righteousness or a zealous evangelism. We all know only too well the religious zealotry of this ilk. It has made and would make community life unbearable. Humanity has yet to develop an institutionalization of the private realm that is adequate for the times. We have over-institutionalized the public realm and filled the public space with what seem never ending bureaucracies. We have under-institutionalalized the private domain and left millions of people swimming in that sea of eccentric and chaotic privacy that lacks order and meaning.[25] White's poetry is a call to apply the new

---

23 T. H. Qualter, Opinion Control in the Democracies, London: Macmillan Press, 1985, p. 23.
24 'Abdu'l-Bahá, op. cit., pp. 111-112.
25 E. Cohen, N. Ben-Yehuda, and J. Aviad, "Recentering the World: The Quest for Elective Centres in the Sociological Universe", Sociological Review, 35, 1987, pp. 320-346.

instrument of Bahá'u'lláh's Order to create communities worth the name community.

No art in any of its varied expressions can be great unless it is timely in terms of need, capacity and response. Artists should not wear their knowledge or talent as arrogance. Worn as humility, the artist tempers his voice and trains his vision with compassion, sensitivity and understanding.

The same is true of genuine community. Bahá'u'lláh's Order provides the moral coordinates of community; it provides an endless series of experiments in creating unity, but the individuals who must apply this increasingly precisioned instrument need to perfect their use of language. White is a model, not of an esoteric poet who plays with words well, but of a way of applying the Bahá'í model in community.

Human beings respond not from the sheer weight of argument and verbal persuasion, not from having their minds infiltrated with clever arguments or manipulation of words. Clever artists can easily titillate or provoke the imagination of others. But the artist will not find resonance in people's hearts and minds unless he establishes a trust and leans forward with his readers to learn. His own vulnerability and weakness tempers his vision and, in the process, the river of people's hearts start to run.[26]

The fine delicacy in human relationships is at the core of White's poetry. He is never embittered, although he often shrinks and is appalled. His poetry, too, is the agony of the creative individual whose nightly rest only resuscitates him so that he can endure his agonies the next day. But there is a golden seam of joy that comes from his passion, his commitment and his realization that he is helping to form the structure, the vocabulary, the sensibility, of a new Order.

White knows that he is working at the beginning of this new world and with all the strangeness, darkness and insecurity that mark all true

26 B. Nahkjavani, op. cit., p. 15.

beginnings.[27] White deals with the existential questions of the human predicament with both timeliness and timelessness. In the process he helps his readers to define who they are, where they've been and where they want to go.

The social, moral and intellectual issues of our times have attained a staggering complexity. The Bahá'í perspective is summoning us "to cease the jangling mockery of our paltry imagination in order to give ear to the universal tolling of this new Revelation"[28] As artists we need a command over the craft to perfect our instruments to their highest ability.

It is in the inner life that this perfection is achieved. The process does and will require all that we have. It also requires a profound appreciation of just what is involved in this inner life and in the affects of this inner life on our environment. An acute exercise of judgement is involved and will be demanded of us in the years ahead.

---

27 Norman K. Swazo, "The Ontological Meaning of the World Order Movement", Alternatives, X, Fall 1984, p. 295.

28 B. Nahkjavani, op. cit., p. 19.

EPILOGUE

Richard W. Bell[1]

Despite embracing a wide range of contributions, the present volume has been selective. In summing up, I will highlight the more important themes to emerge from the papers and from the Symposium at which they were first presented.and draw attention to the important gaps in the present volume.

The most glaring omission is that there has been no formal contribution from women which is barely excusable given that the Bahá'í perspective on the environment ascribes an important role to the feminine perspective of the environment. [2] It postulates that equal participation by women in decision-making about natural resources will promote environmental care. Women bring to decision-making processes those attributes in which they are strong such as mental alertness, intuition, and the spiritual qualities of love and service. Without that input, decision-making about natural resources will continue to be dominated by the forcefulness and aggressiveness that are more characteristic of men.

Indigenous peoples have an important perspective on the environment. By contrast with Western society which has tended to act as if it lives in a world that has no limits, many, if not most indigenous communities had some sense of living within an environment subject to real limits. As we strive to re-instill that realization and its implications for our behaviour and consumption in a global community, we can perhaps learn some valuable lessons from the Aboriginal people of Australia and

---

1 Lecturer in Environmental Science, School of Biological and Environmental Sciences, Murdoch University, Murdoch, W.A. 6150.

2 Bahá'í International Community, The Bahá'í Statement on Nature, New York: Bahá'í International Community, 1987, p. 7.

elsewhere in the world. Dr. Meyer alluded to that in his paper, but the theme could have been explored more thoroughly.

In future discussions of this sort, the point of view of business should be represented: they have a perspective that needs to be heard in the debate about the environment. It would be wrong to characterize all business people as acting out of greed. Many are motivated in their business ventures by a desire to create wealth in society so as to enrich the society through better education and richer culture. In relation to economic development, the key issue is how to ensure that production of goods and services accurately reflects their real costs in terms of environmental damage. Currently many products are financially lucrative for manufacturers but may be economically unviable because many of the real costs of production are externalized. For example, air pollution caused by industry is a cost which is often borne by the community in terms of acid rain, forest decline, and respiratory disorders etc.

The authors have alluded to at least two risks in discussions on the environment. The first is that they will be characterised by a short burst of enthusiasm that is replaced by apathy. A serious challenge for the proponents of a sustainable future is to organize and to capture that spirit of enthusiasm and to sustain it before it is replaced by apathy. Dr. Newman in his paper and other authors have stressed the need to cultivate hope and to be positive. This sense of optimism and hope is a most important ingredient in decision-making about the environment. Environmental activists who are genuinely optimistic create the real possibility of changing things and accomplishing their goals for a sustainable future. By contrast, those who are overcome by despair or apathy, are not going to accomplish their objectives. A more serious concern is that environmental activists, the agents of change, are often deeply pessimistic about the prospects for positive change because of a mistaken view of the nature of human beings as incorrigibly aggressive. This paradox where the agents of change lack a deep conviction about the efficacy of their labours was pointed out previously in relation to achieving World

Peace: it has the potential to paralyze the entire movement for a sustainable future.[3]

The notion of service was not discussed to any great extent in the present papers. Inevitably, most changes projected for a sustainable future require genuine and probably sustained sacrifice in the service of our fellow humans. For example, readjusting the balance between rich and poor nations requires us to have a deeply felt sense of them as being part of our family and being prepared to make some sacrifices in order to enhance their quality of life. In nurturing both hope and service, it is difficult to ignore the importance of true religious faith for the peoples of the world. Indeed, the level of service, sacrifice and hope required may well be impossible without true religious faith as suggested by 'Abdu'l-Bahá.[4]

The second risk is that the discussions on the magnitude of the world's environmental problems are going to degenerate into a polarised debate about who is the cause of the problem. There is already a potential for that. Many developing countries are very resistant to the messages that are coming out of the developed countries to the effect that it is actions in the Third World such as rainforest clearing which are damaging our environment. Third World countries take the view that they have a primary responsibility to alleviate poverty and uplift the standard of living for their people and they are not pleased to be told by the rich nations that they should forego these minimum benefits and they alone should make the sacrifices for the future of the planet.

In the present volume, the need to develop better conflict resolution and negotiation skills has not been mentioned to any great extent. Yet authors have proposed bringing together diverse groups, and raised complex problems that involve relationships among diverse and some-

---

3 Universal House of Justice, The Promise of World Peace, Mona Vale: Bahá'í Publications Australia, 1985, p. 22.

4 'Abdu'l-Bahá, Secret of Divine Civilization, Wilmette: Bahá'í Publishing Trust, 1975.

times discordant countries, among groups of professionals, and among community groups. There is a risk that we do not have adequate skills to bring all those people together to talk in a sensible way in order to reach agreed solutions. Skills in negotiation, conflict resolution and values clarification are an important part of this process that we have not explored properly.

Change has been a recurring theme of the the papers in the present volume. Most authors have foreshadowed substantial changes in relationships between nations, between regions within nations within communities and among individuals.

For the individual, changes seem to be predicated on the need to understand better the way in which human individuals are related to the environment and to one another. We need to educate ourselves and our children to a full appreciation and awareness of these inter-relationships: the fact that we are not isolated, that we are not disconnected, that we are part of a global environment. The quotation below encapsulates the notion of interrelationships and was chosen as the theme of the symposium on which the present volume was based.

> We cannot segregate the human heart from the environment outside us and say that once one of these is reformed everything will be improved. Man is organic with the world. His inner life moulds the environment and is itself also deeply affected by it. The one acts upon the other and every abiding change in the life of man is the result of these mutual reactions. [5]

Institutional changes are required at the international and national level. They also clearly involve interrelatedness. Dr. Chittleborough pointed to the fact that the growth of cities, for example, is alienating valuable agricultural land and drew attention to the need for planners and

5 Shoghi Effendi, In: A Bahá'í Perspective on Nature and the Environment, Mona Vale: Bahá'í Publications Australia, 1989, p. 10.

governments to protect valuable agricultural land from alienation for other purposes. Dr. Moshirian outlined the impact of trade imbalances on the survival of the forest. Despite the very complex interrelationships involved in these issues, I was very encouraged by his synopsis that we can take hope from the creativeness of human beings and the possibility that out of a potential crisis, the innovation of human beings will come to the fore. Although generally portrayed as a potential disaster for humanity, environmental degradation can become for us the stimulus to work together, to use our inventiveness and to come up with ideas, products and solutions of which we have never dreamt before.

Finally, there is an urgent need for change at the local level. Barry Oldfield reminded us that, we have be prepared to take action, to be involved and to implement the solutions. Barry Oldfield advocated a range of tree planting activities in which individuals can be involved. Phil Gabriel discussed other simple community based activities in which we can be involved. For local action, there is a dearth of workable models of communities that are able to implement these principles, live by them and demonstrate their capacity to solve the problems that face communities. It is always much easier to make change when you can examine workable models. Parents and teachers are aware of the power of models for the acquisition of skills, attitudes and behaviours in children. The same may be true of whole communities. The challenge to communities such as the Bahá'í world community is to be visible as models for change, as models for how communities can inculcate the spiritual values of hope, service, love, sacrifice, worldmindedness and sincerity, how they can allow individuals to grow and in encouraging individuals to grow to present a positive vision which enables communities to take control of their destiny and to invent a sustainable future.

Further Reading

I. Dankelman and J. Davidson, Women and environment in The Third World: alliance for the future, London: Earthscan Publications, 1988.

J. Beanclerk, J. Narby and J. Townsend, Indigenous peoples: a field guide for development, Development Guidelines 2, Oxford: Oxfam, 1988.

BIOGRAPHICAL INFORMATION ON AUTHORS

Richard W.Bell

Lecturer in Environmental Science, Murdoch University. Dr Bell has lectured in Soils and Land Use in Fiji, and conducted research on food legumes in Queensland, Western Australia and Thailand. He has particular interest in Rural Development, Research and Training for Agricultural Development and Sustainable Agricultural Systems. He has been a Bahá'í since 1975, a member of Bahá'í communities in Fiji, Brisbane and Fremantle and was recently appointed as an Environmental Consultant to the Bahá'í International Community's Office of the Environment.

David Chittleborough

Lecturer in Soil Science at the Waite Agricultural Research Institute, University of Adelaide. His research interests are in Soil formation and distribution, soil pollution and the impact of urbanization on agricultural land. He has served on the National Spiritual Assembly of the Bahá'ís of Australia for several years and is its former Chairman.

Phillip Gabriel

Former Research officer with Friends of the Earth. Friends of the Earth are an international group interested in and committed to making a difference in environmental awareness and management by the entire community. A graduate in ecology from Murdoch University, Mr Gabriel has worked on Wetlands research and is involved in a number of W.A. conservation groups.

Phillip Jennings

Associate Professor of Physics at Murdoch University. For the past 15 years he has carried out teaching and research in several areas of physics, including radiation physics, solar energy conversion, environmental and surface physics. He is active in public life as a speaker, consultant and activist. In recent years his major community activities have been radiation protection in mining, wetlands conservation, science and technology policy and renewable energy policy. He is currently President of the Conservation Council of WA and the Wetlands Conservation Society.

Ray Meyer

Former Head of the Behavioural Sciences Department of the Goulburn College of Advanced Education. His doctoral thesis was awarded with distinction from the University of California, Los Angeles. During this period he studied the effects of the environment on the perceptions and performances of students in universities and colleges and pioneered concepts for the development of Colleges of Advanced Education in Australia. For his contribution to opening up the field of Health Education in Australia and in developing curricula for Mankind, Schools and Worldmindedness, he was awarded a fellowship of the Australian College of Education. Since retirement, Dr Meyer has devoted his time to the administration of the Baha'i Faith, serving as a member of the National Assembly of the Bahá'ís of Australia, as its secretary and as a member of its national committees. His concern for the environment arises from his friendship with Aboriginal people, the immediate threat to existence from nuclear warfare, greed and a seriously disturbed ecology.

Fariborz Moshirian

Lecturer in International Economics at the University of New South Wales. His specialises in applications of International Economics to fields such as International Trade and International Banking and Finance. Dr Moshirian is developing a General Equilibrium Model for World Economy in which the environment is one component and his research in the field of Development Economics consists of application of the New International Economic Order to environmental effects of economic growth.

Peter Newman

Director of the Institute for Science and Technology Policy, Murdoch University. Lectured at Murdoch University in Environmental Science, 1974-1992. Has served as Director of Environmental Planning and Development, Office of the Cabinet in the Western Australian Government, as Advisor to the WA Minister for Transport and Regional Development (1985-86), as a member of the Environmental Protection Authority (1987-89), and as a Councillor of Fremantle City Council (1976-1980). Professional interests in Environmental Planning Policy and Urban Transport.

Barry Oldfield

Filmmaker whose work since the mid-1970's has concentrated on environmental issues, in particular the need for massive environmental correction through tree planting and care of forests.

Films: On the Edge of the Forest with EF Schumacher

St Barbe of the Trees with Richard St Barbe Baker

Wheat Today, What Tomorrow with David Bellamy

Has farmed or practised forest management for 28 years. Since 1984, he has worked in a full time voluntary capacity to promote planetary care through individual responsibility. Foundation member and since 1987 WA State President of the The Men of the Trees.

Ron Price

Lecturer in Adult Education, Thornlie College of Technical and Further Education. Mr Price has taught in tertiary institutions, colleges of Technical and Further Education and schools, and worked as a journalist. His main area of expertise is in personnel management, communications and industrial relations. He is currently working on a Masters degree in the Sociology of Religion.

Jimmy Seow

Senior Environmental Officer with the Petroleum Division of the WA Department of Mines, and completing a Doctor of Philosophy degree at the University of Western Australia in the field of integrated catchment management. He is currently Executive Secretary of the Petroleum Liaison Committee and member of the Technical Advisory Committee of the State Combat Committee for Combatting Marine Oil Pollution. He has worked as a geophysical consultant in the oil and gas exploration sector in various parts of the world and as a geologist in Australia. He has been a Bahá'í since 1970 and has lectured on various Bahá'í subjects in different parts of the world.

Keith Suter

Executive Director, National Goals and Directions Movement, and previously Director of the Trinity Peach Research Institute, Perth. He is well known in Australia for his involvement as a writer, lecturer, broadcaster and activist for Peace. He has been actively involved with the United Nations Association of Australia and has been its WA State President and a past national President of the Association. He has a major interest in international relations and is the author of several books on the subject including "Reshaping the Global Agenda - The United Nations at Forty".

Bruce Wilson

Currently in the Information Technology Department of the WA Water Authority. Bruce has been actively involved in Community Aid Abroad(CAA) for 10 years and in that time has been President of its WA executive, and a member of the National executive. He currently is a member of the WA executive of CAA.

A Bahá'í View on the Environment

> Nature in its essence is the embodiment of my name, the Maker, the Creator. Its manifestations are diversified by varying causes, and in this diversity there are signs for men of discernment. Nature is God's Will and is its expression in and through the contingent world. It is a dispensation of providence ordained by the Ordainer, the All-Wise. It is endowed with a power whose reality men of learning fail to grasp.
>
> Bahá'í Writings

With these words, Bahá'u'lláh, Prophet-Founder of the Bahá'í Faith, outlines the essential relationship between man and the environment: that the grandeur and diversity of the natural world are purposeful reflections of the majesty and bounty of God. For Bahá'ís there follows an implicit understanding that nature is to be respected and protected, as a divine trust for which we are answerable.

The Bahá'í Faith's theology of unity and interdependence relates specifically to environmental issues. Again, to quote the Bahá'í Sacred Writings:

> Liken the world of existence to the temple of man. All the organs of the human body assist one another, therefore life continues...Likewise among the parts of existence there is a wonderful connection and interchange of forces which is the cause of life of the world and the continuation of these countless phenomena.

A century ago, Bahá'u'lláh proclaimed that humanity has entered a new age. Promised by all the religious Messengers of the past, this new epoch will ultimately bring peace and enlightenment to humanity. To reach that point, however, humankind must first recognize its fundamental unity - as well as the unity of God and of religion. Until there is a general recognition of this wholeness and interdependence, humanity's problems will only worsen.

"The well-being of mankind, its peace and security, are unattainable unless and until its unity is firmly established," Bahá'u'lláh wrote. "The earth is but one country and mankind its citizens."

The major issues facing the environment movement today hinge on this point. The problem of ocean pollution, the extinction of species, acid rain and deforestation - not to mention the ultimate scourge of nuclear war - respect no boundaries. All require a transnational approach.

While all religious traditions point to the kind of cooperation and harmony that will indeed be necessary to curb these threats, the religious writings of the Bahá'í Faith also contain an explicit prescription for the kind of new world order that offers the only long-term solution to such problems.

Built around the idea of a commonwealth of nations, with an international parliament and executive to carry out its will, such a new political order must also, according to the Bahá'í teachings, be based on the principles of economic justice, equality between the races, equal rights for women and men and universal education.

For example, the uplifting of women to full equality with men can also help the environmental cause of bringing a new spirit of feminine values into the decision-making about natural resources. The scriptures of the Bahá'í Faith note that "man has dominated over woman by reason of his more forceful and aggressive qualities of both body and mind. But the balance is already shifting; force is losing its dominance, and mental alertness, intuition and the spiritual qualities of love and service, in which women are strong, are gaining ascendancy. Hence the new age will be less masculine and more permeated with feminine ideals..."

The Bahá'í Faith seeks above all else to revitalise the human spirit and break down the barriers that limit fruitful and harmonious cooperation among men and women, whatever their national, racial and religious background.

For Bahá'ís the goal is to carry forward an ever advancing civilisation. Such a civilisation can only be built on an earth that can sustain itself. The Bahá'í commitment to the environment is fundamental to our faith.

Bahá'í Declaration prepared by the Environment desk of the Saskatoon Bahá'í community.